D0837448

THE BIG BANG TO NOW

ALL OF TIME

IN SIX CHUNKS

TERRY HERMAN SISSONS, PH.D.

WWW.ALLOFTIMEONLINE.COM

The Big Bang to Now: All of Time in Six Chunks

Copyright@T.Herman Sissons, Ph.D. 2012
All rights reserved

ISBN: 13:978-1477638507

Library of Congress Control Number: 2102912097

Cover Photograph: Credit Apollo 12 Crew NASA

Printed by CreateSpace
North Charleston, South Carolina

To order additional paperback or e-copies of
The Big Bang to Now: All of Time in Six Chunks
go to one of the following:

www.AllOfTimeOnLine.com
www.Amazon.com
www.Amazon.co.uk
www.Amazon.de
www.Amazon.fr
www.Amazon.es
www.Amazon.it
https://CreateSpace.com/3896348

DEDICATION

This book is dedicated to anyone who has

ever looked at the stars and wondered

how they came to be.

And where we are all going.

ACKNOWLEDGEMENTS

Thank you to those scholars who have shared their enthusiasm along with their theories and discoveries about this great universe of time. Along with my profound gratitude, I offer my apologies. I have reduced years of dedicated expertise and hundreds of books and articles to less than a page, sometimes even to a mere phrase. They deserve more than that, and I hope my few words will serve not as substitutes but as signposts to excite others to learn more.

I also want to thank all those who read the first edition of **The Big Bang to Now** and told me what they thought. The feedback, the enthusiasm, and above all, the questions, have contributed to the shape and content of this second edition in ways that, left to my own devices, I could not have foreseen.

My special gratitude for Karen Peterlin who has read my words with perhaps more careful attention than they deserve, and who has saved me the embarrassment of parading my mis-spellings, typographical and formatting errors, and some sheer incomprehensible babbling.

Once again, thank you to Peter Sissons. How much of the best in this book is due to his support is unquantifiable.

CONTENTS

DATING CONVENTIONS

In order not to complicate the understanding of time further, this book uses everyday terms as much as possible. *Bya* refers to *billions of years ago*, *mya* to *millions of years ago*, and *ya* to *years ago*.

Various dating systems are used in various fields of scientific study, however, and it is helpful to understand them when they are used in other sources.

BP (Before Present) represents the total number of years before now.

BCE (Before the Common Era) dating is equivalent to BC (Before Christ) dates and is also equal to BP dates plus 2,000 years. Thus, 5,000 BCE is equal to 5,000 BC, to 7,000 BP, to 5,000 AD, and to 7,000 years ago.

AD (After Christ) or CE (Common Era) refer to events occurring in the last 2,000 years. It is the dating convention used in most personal and business notations today.

GEOLOGICAL AGES

Geological ages are the official time line of Earth used by scientists. They are based on findings of rock and early fossils and are the best estimates we have of what happened when. In order to make it easier to learn more about any particular event, the geological age in which it occurred is noted following each event. The full geological time line is also included in the appendix.

Preface

The Story of Time

The story of time is amazing. It's fantastic, almost incredible. For most of us, it's mind-boggling.

And that is the problem for many of us. We hear numbers attached to years that sound so huge that they become meaningless. A billion years sounds just as incomprehensibly long as a hundred million years even though it's ten times as long. Ten billion years might sound barely more than a hundred thousand years, even though it's ten thousand times longer. Everything just sounds like it happened a very long time ago.

But it is possible for quite ordinary people whose mathematical abilities do not extend too far beyond multiplication and long division to understand all of time.

It doesn't require algorithms or complex equations. It doesn't require knowing almost everything that's ever happened. If someone can memorize a telephone number, knows their social security number by heart or even just a zip code, they can begin to understand all of time.

It might take a little bit of effort. But it's a lot easier than filing a tax return or maybe even than keeping track of a bank account.

This book is about how to do it. It's about how to understand all of time.

There are many ways to tell the story of time. The story here is the way science tells the story today. It's not a fairy tale, even though parts of it probably aren't true. Scientists make a lot of mistakes along the way, and as our mechanisms for peering beyond the present improve, the story will change too.

It is important, however, to understand that although this is the story of time as it is currently told by science, nothing in it, including the Big Bang and the theory of evolution, is based on the assumption that there is no God.

Science is not a tool that can be used to prove whether or not there is a God. Science does, however, show us much about the world that we cannot see in any other way. It can, therefore, tell us that some anthropomorphic ideas of God are based on views of the world that no longer make sense. Religious beliefs that do not take into account the discoveries of science, may reflect a denial of the experiences of life and of the universe through which believers argue divinity is revealed.

Science can challenge us to examine our beliefs. It can broaden and deepen those beliefs, or lead us to dismiss some as based on a view of the Universe that is no longer tenable. But science neither proves nor disproves the existence of God, nor can it answer many of the questions often addressed by religion. It is not science therefore, that determines whether someone is a believer.

How we interpret our experiences and whether they give us a glimpse of what many believers term "God" has been the subject of great controversy down the centuries.

What one ultimately decides about questions traditionally answered by various religious teachings go beyond the scope of science.

Introduction

How to Think about Time

The Universe and time as we know it started almost 14 billion years ago. The dinosaurs became extinct a mere 65 million years ago. But a million or a hundred million, a hundred thousand or even just twenty-five thousand years are all so long ago that for many of us, these vast amounts of time melt into a shapeless blur.

If ten million years sounds very much like a hundred million years which doesn't sound too different from ten billion years, don't worry. There is an easy and fast way to get over this seemingly gigantic hurdle.

Chunking

The trick to understanding mega periods of time is to chunk. Usually, people can't remember more than 5-7 items at a time. Unless we chunk. That's how we remember telephone numbers. By grouping separate numbers into chunks like access and area codes, local numbers and extensions we often remember 20 numbers or even more.

Chunking dates works just as well as chunking telephone numbers.

That's why this book is divided into six sections or chunks.

Learn just one Mega-Chunk

This book is based on six big chunks, each with up to 25 turning points in the history of time. They are all fascinating and important.

But knowing just six or seven anchoring events can transform your understanding of time. Memorize the following seven events and see what happens immediately to your understanding of all of time.

- The Universe began with the Big Bang
  ...about 14 bya

- Life appeared on Earth shortly after Earth formed
  ...about 4.5 bya

- The dinosaurs evolved 250 mya;
  they died out 65 mya

- The first humans, ancestors of *Homo sapiens*, evolved
     ...about 1.8 mya

- *Homo sapiens* began in Africa
  ...about 250,000 ya

- Settled farming and the first civilizations began
  ...about 10,000 ya

- The scientific revolution began
  ...in the 1400's AD

Finally

On the last page of this book is a short outline of all of time. Cut it out or print it and use it as a book mark to keep oriented as you read this book..

Reminder

③ A billion years is a thousand times longer than a million years.

③ A million years is a thousand times longer than a thousand years.

Chunk I

Five Big Start-Ups

Billions of Years Ago

Five big beginnings and their dates anchor all the other events that took place between almost 14 and just a quarter billion - or 250 million - years ago. Earth was formed when the Universe was already nine billion years old. Micro biotic life began on Earth almost immediately and thrived for more than two billion years.

After the Great Oxygen Poisoning, one of the greatest extinction events Earth has ever witnessed, life took a dramatic turn. New and more complex life forms evolved, first of plants and then of animals in the sea.

Life then moved onto land, a procession led by plants and eventually followed by animals. This period was punctuated by several mass extinctions, and ends with a massive extinction that, like extinctions before, resulted in another dramatic change in life on Earth.

1. ✳ 13.7 billion years ago: The <u>Big Bang</u> started it all when our Universe and time and space as we know it began.

2. ● 4.6 billion years ago: <u>Earth</u> began when an arm of the Milky Way collapsed, creating our solar system.

3. ❋ 3.9 billion years ago: <u>Life</u>, tiny organisms called microbes, appeared on Earth quite a short time after Earth itself was formed.

4. ♠ 1.7 billion years ago: A little than more two billion years after life was first established on Earth, or 1,700 million years ago, the first <u>true plants and animals</u> evolved.

5. 🐌 470 million years ago: Just under a half billion years ago, the ozone layer was developed enough to protect plants and animals from the sun's rays outside their watery habitat and <u>life on land</u> began.

GEOLOGICAL TIMES IN THIS CHUNK
EON: HADEAN................4.55 ------------------ 4.00 BYA
ARCHEAN...............4.00 -------------------- 2.5 BYA
PROTEROZOIC..........2.5 BYA ---------------- 542 MYA
PHANEROZOIC..........542 MYA ------------- PRESENT
ERA: PALEOZOIC..............542 --------- 251 MYA
PERIOD: CAMBRIAN...........542 - 488 MYA
ORDOVICIAN..........488 - 444 MYA
SILURIAN.............444 - 416 MYA
DEVONIAN............416 - 359 MYA
CARBONIFEROUS....359 - 299 MYA
PERMIAN.............299 - 250 MYA
(See Appendix for a full geological time scale)

Chapter 1

The Universe Begins

13.7 to 4.6 Billion Years Ago

Five Big Start Ups: *Billions of Years Ago*

Time Before Earth

This is the beginning of the story of how you and I have come to be here. From the small porthole of human experience, whatever happened over 13 billion years ago when time actually started might seem just a little irrelevant. Yet the amazing thing about time is that everything that's ever happened in all this vast time is connected.

Step by step, one thing follows another with a kind of wonderful irrevocable beauty.

Where we are now started a very long time ago with a Big Bang.

The Big Bang

13.7 billion years ago

Astro-physicists estimate that time and space began about 13.7 billion or 13,700 million years ago. Everything in the Universe is made from energy which, as far as we can tell, cannot be either created or destroyed. In this sense, energy is eternal.

There may be a whole prehistory to the Universe before the Big Bang but for now, we begin our time, Time Zero, when, in less than 3 minutes, 98% of all matter that exists today exploded in a great fireball into the space it created as it inflated from an unimaginably intense concentration of energy smaller than a dot. The dot is called a *singularity*, but although it has a name, there is more we still don't understand about it than we do.

4% of the Universe is made of the ordinary matter we usually think of as "everything." But to make sense of what we do perceive, scientists think about 23% of the Universe must be dark matter which we may potentially be able to detect in special high-intensity conditions such as a Haldron Collider. Another whopping 73% of the Universe seems to be something called "dark energy." Most scientists today think it must be there but they have no idea what it is. Some scientists think, however, there isn't such a thing as dark energy at all and that we are misinterpreting the data.

Right from the start, you can see how much we don't know.

Five Big Start Ups: *Billions of Years Ago*

more about…
the big bang: unanswered questions

It's hard to understand the theories about the Big Bang if you aren't a physicist, so many people don't realize how much even scientists don't know about how our Universe began. Although scientists are making educated guesses, nobody knows for sure:

● whether the singularity, that dot of energy that exploded into the Universe, always existed, or if it didn't, where it came from.

● what made – or makes - the energy explode into time and space when it did.

● if the Universe will eventually reach equilibrium, if it will keep expanding forever, or if, even now, galaxies are expanding, slowing down, or moving in a flow toward a single concentrated point. None of these possibilities are unreasonable in the light of what we know.

● if this was the first Big Bang, or if the expansion and contraction of the Universe has occurred before and will keep on happening forever.

We also don't know:

● if there are other Universes, and if there are, whether they operate under the same laws of physics as our Universe.

● what might happen if another Universe collided with ours.

It seems the more we learn, the bigger the questions become.

Chapter 1: The Universe Begins - *Time Before Earth*

Simple Atoms

13.69 billion years ago

For the first 300,000 years after the Big Bang, the Universe was indescribably hot but as it expanded, it cooled enough for particles of matter with opposing electrical charges to attract each other to form simple atoms. When all the electrons were attached in simple atoms, the primal radiation began to spread throughout space.

Today this radiation is detectable to us on Earth. Among other things, it is the cause of about 1% of the "snow" on an untuned TV.

Five Big Start Ups: *Billions of Years Ago*

more about…
particles, atoms and molecules

The tiniest particles of matter that began to combine with each other after the Big Bang come in a dazzling array.

● Elementary particles include <u>Quarks</u>, which come in six "flavors" and <u>Leptons</u>, which include *electrons*. There are also 12 particles called <u>Force Carriers</u> which include *gluons*, *bosons* and *photons*. These elementary particles combine to make composite particles like <u>Protons</u> and <u>Neutrons</u>.

● Atoms are made of a nucleus packed with positively charged protons and neutral neutrons, with negatively-charged electrons orbiting around the nucleus. They are sub-microscopically small, but relatively speaking, there is a lot of empty space in an atom.

There are 92 different kinds of naturally-occurring atoms made of varying numbers of protons, neutrons, and matching electrons. Simple atoms formed when the Universe first started. The other 90 atoms still form in the intense heat of stars.

● When atoms combine with each other, they are called molecules. Almost everything in our everyday world is made up of these combinations.

At first, we thought the Universe was simply made up of these atomic building blocks. But then scientists stumbled on two big questions. The first Is the discovery that the laws of space and time on the molecular level seem totally different from the way space and time appear to operate in the larger world. Resolving the conflict between what is called The Standard Model that describes the sub-atomic world, and Einstein's Theory of Relativity describing the larger world is still a scientific challenge.

The second question is why matter and anti-matter did not destroy each other on contact immediately after the Big Bang. That that isn't what happened is really quite wonderful. Because all of us and everything we can see is made of the magnificent stuff we call matter.

Chapter 1: The Universe Begins - *Time Before Earth*

Stars

13.7 billion years ago

The Universe existed in cosmic darkness for more than 100 million years. Light appeared with the first stars.

Stars have always done more than twinkle. Besides creating light, the stars formed the Universe as we know it. They first manufactured and then dispersed heavy elements, a necessary step to the formation of solar systems like ours. They created the black holes that form the hearts of galaxies. It is a star we call the sun that is the source of all the energy that supports life as we know it on Earth.

It is in the heart of stars that the elements of which we ourselves are made were first produced. Each of us is quite literally made of star dust from explosions at least 5 billion years ago. If it weren't for stars, we wouldn't be here. Or if we were, we would be made out of something completely different than carbon, nitrogen, oxygen, and a good deal of everything else generated by stars and out of which we and our Earth are made.

Five Big Start Ups: *Billions of Years Ago*

Stars come in a mind-boggling array of sizes, shapes, and colors. There are brown and white dwarfs, blue and red giants, baby stars and twin stars.

Proto stars or baby stars begin within a galaxy where clouds of gas and dust are squeezed together by the force of gravity that creates a nuclear fusion reactor at its heart. During most of its life, a star burns steadily, and those we can see often give us direction in the night sky.

A star may burn for billions of years, but eventually it starts to run out of fuel and collapse. This generates intense heat making expiring stars the great creative engines of the Universe, continuously remaking, renewing and recycling everything in it. Depending on their size, stars eventually die as white dwarfs, novae or supernovae. The collapse of some of the biggest stars create black holes.

As they die, stars throw out clouds of "star dust," or more prosaically, cosmic debris. Some of this dust becomes planets, including the rocks of our own Earth. Some star dust is forged into the oxygen, carbon, nitrogen, and other essential components such as water out of which each of us is made and that make our world the kind of place it is.

Chapter 1: The Universe Begins - *Time Before Earth*

Galaxies

13.2 billion years ago

As the hydrogen and helium gas formed, it expanded and, under the force of gravity, gathered into billions of great clouds. These clouds formed into huge clusters with gigantic empty spaces in between. Within 500 million years after the Big Bang, galaxies had begun to form. It is the time when the Milky Way, the galaxy where we now live, first began.

Today the new galaxies are no longer forming, but galaxies are continuing to move apart from each other. That means that Earth will not get any further away from other stars and planets within the Milky Way. But since the Universe is continuing to expand, stars and solar systems in the millions of other galaxies are getting further and further away every day.

Five Big Start Ups: *Billions of Years Ago*

The size of the Universe is incredible and it is still getting bigger. It is dotted with galaxy clusters which are separated by almost incomprehensibly huge bubbles of apparently empty space as big as a billion light years. There are at least 100 billion galaxies each with millions, billions, even a trillion, stars.

In this great vastness, the galaxy we know best, the Milky Way where we live, measures a hundred thousand light years across. We can see part of it spread out – a starry road above us – on a clear night. Our nearest galaxy neighbor, Andromeda, is a measly 2.5 million light years down the road.

How far is that? A light year is the distance light travels in a year. Light travels at 186,000 miles, or 300,000 kilometers, a *second* which makes a light year more than 5 trillion miles. Andromeda is a short trip of 15 trillion miles or so. To put this in perspective, the drive from New York to California is about 2500 miles.

The size of outer space is barely conceivable, and mere words like millions and billions don't convey to most of us the overwhelming immensity of the Universe.

Chapter 1: The Universe Begins - *Time Before Earth*

Chapter 2

Our Earth

4.5 to 4 Billion Years Ago

Five Big Start Ups: *Billions of Years Ago*

Born in Fire

For its first 100 million years, Earth was a ferocious cauldron completely inhospitable to even the most intrepid and adaptable forms of life. Objects hurling around space smashed into it, rain was corrosive acid, and the atmosphere was a ghastly mix of sulphurous fumes.

But our planet would emerge from this terrible childhood as a young nurturing adult, with the capacity to become our Mother Earth.

By the end of this period, Earth had a companion moon, land had begun to form, and, crucially, water had begun to pool on Earth's surface.

Our Sun and Planets

4.56 billion years ago

9 billion years after the Big Bang, our solar system began in an unusually dense swirl of star dust. One of the spiral arms of the Milky Way collapsed under the weight of gravity and grains of dust stuck together into larger and larger clumps as they bumped into each other. 99.9 percent of the swirl eventually became the sun. The leftover .1 percent went into the planets and other objects in the sun's orbit.

Our solar system took about 2 million years to get started. Earth grew from a few grains to a baby planet in 20-30 thousand years, and in 200 million years was essentially the size it is today. As we shall see, it did not have a peaceful start.

Hadean Eon.....4.55 – 4.0 bya

Five Big Start Ups: *Billions of Years Ago*

more about…
our solar system

Eight bodies orbiting around the sun have full planetary status.

The First Four Rocky Planets
● *Mercury* is small, closest to the sun and so the hottest.
● *Venus* is called our Morning and Evening Star. It's about the same size as Earth but scorching hot with an acid atmosphere.
● *Earth*, the next planet, is perfectly placed to be our home.
● *Mars,* the Red Planet, may have water and hosted microbiotic life.

The Next Four Gaseous Planets
● *Jupiter*, with rings and 28 moons, is a thousand times bigger than Earth. Its gravity makes it our big brother, taking many of the blows from objects that would otherwise bombard our home planet.
● *Saturn* has spectacular rings and a moon named Titan which may resemble Earth as it was 4 billion years ago.
● *Uranus*, the third largest planet, has 22 moons, and a blue-green methane atmosphere.
● *Neptune* is 60 times the size of Earth. It has four rings, a blue methane gas exterior and planetary winds over 1200 mph.

A Dwarf Planet, Dwarf Galaxies and Beyond
● *Pluto* is a little ice ball, a maverick we didn't know about until 1930. It circles in an erratic orbit and is so small it was demoted in 2006 from its status as a planet to that of a Dwarf Planet. It is, in other words, the biggest of the little guys.

There are also swarms of sub-planet sized objects swirling around the outer edges of our solar system. The first group belong to the *Kuiper Belt*, the second group contains *Scattered Disk* objects, and the third group, the *Oort Cloud*, is probably the source of most of the comets we see flashing through our solar system.

Beyond that dwarf galaxies are orbiting the Milky Way, which may be made of that mysterious dark matter scientists are puzzling over.

Chapter 2: Our Earth - *Born in Fire*

Hell Fire on Earth

4.55 billion years ago

Once it was big enough to be called a proper planet, Earth underwent a harrowing baptism. Its first half billion years are called the Hadean Eon, and this time does sound like a time out of hell. It blazed as a ball of fire for 100 million years and for another 400 million years was bombarded with spatial rubbish left from the formation of the solar system.

The atmosphere was a lethal mixture of methane, hydrogen sulfide and carbon dioxide. The rain was caustic and volcanoes hurled torrents of debris everywhere. Light from a weak sun was blocked by dense clouds. There was no oxygen at all.

It really was quite seriously awful.

Hadean Eon.....4.5 - 4.0 bya

Five Big Start Ups: *Billions of Years Ago*

more about…
our wondrous planet

We are only now realizing what a special home Earth is for us.

● Earth is just the right distance from the sun: 5% closer or 15% further away and our water would either boil or freeze.

● Our Sun is just the right size: if it were much bigger, it could have burned itself out in 10 million years.

● The Milky Way was fairly old by the time our solar system formed in a part of the Universe unusually dense with matter. This made the area rich with the molecules we need for life.

● Earth has a radioactive core that warms and protects us, and fires the volcanoes that keep us on dry land.

● Earth is in the company of bigger planets. Massive Jupiter, in particular, shields us from some of the huge meteors that could make life unsustainable if they crashed into us instead.

Earth may be unique in the Universe. Scientists think it may be possible to make Mars habitable for humans some day, but that will require overcoming some huge obstacles we haven't negotiated yet.

We also know about at least ten planets in the Milky Way where liquid water might exist and so where human life might be able to survive. Unfortunately all these potential planets are hundreds of thousands of light years away. The best available space ship we have today would take 1200 years to travel <u>one</u> light year. At current speeds, we couldn't get to even the nearest planet we've identified so far in less than 720,000 years. If we did manage the trip and somebody is there before us, we might not be welcome.

So if we want to survive, we had better take care of our Earth. Living somewhere else that doesn't depend on support from Earth isn't going to be possible any time soon.

Chapter 2: Our Earth - *Born in Fire*

The Moon

4.5 billion years ago

Had we been here on Earth, the crash would have felt like an unmitigated catastrophe. When it was about 15 million years old, Earth was hit by a rocky something as big as a medium-sized planet named *Thea*. Huge pieces of Earth were gouged out and flung into space. Pressured and crushed by gravity, this mass of Earth and cosmic fragments became – not an errant catastrophe – but our moon.

Today this fortuitous fellow traveller is our night light, a loyal companion that stops Earth from wobbling too erratically, moderates our weather, and gives us the ocean tides.

Hadean Eon.....4.5 - 4.0 bya

Five Big Start Ups: *Billions of Years Ago*

more about…
pointing north

Besides the moon, Earth has another treasure. The needle of a compass always points north because there is a huge magnet at Earth's center.

> It was formed when the heat was so intense it created a swirling soup of melted rubble. In this primitive brew, heavier metals, including iron, gradually sank, giving Earth a magnetic center that is still burning. It may even be a nuclear reactor, though we haven't been able to drill past Earth's surface yet to find out.
>
> Scientists call this meltdown "the iron catastrophe." But for the prospects of life here, it was another amazingly fortuitous event in the formation of our home. Without this giant magnet, we might not be able to use our compass to head north. More critically, it is this magnet that deflects the solar wind around our planet, keeping our atmosphere from being blown away and protecting us from many dangerous cosmic rays.
>
> The magnetic field is constantly moving, which is why geographical north is not exactly the same place as magnetic north. Between every 20 thousand and 37 million years it also reverses, and the magnetic needle then points south until the next reversal. We don't know why. We also don't know why the strength of the magnetic field waxes and wanes. Right now, it is 35% weaker than it was 65 million years ago. This isn't really great news if you hold your shopping list onto your refrigerator with magnets.

Chapter 2: Our Earth - *Born in Fire*

Land and Sea

4.4 billion years ago

Earth did not begin with solid land but with an unstable surface of molten liquid and ice. A tiny zircon crystal no more than two hairs' width found in Australia is the oldest piece of Earth's crust ever discovered.

The formation of the crust started fairly early in the Hadean Eon, and continued to form in the next age or eon given the Greek name "Archean," meaning, "ancient time."

Hadean Eon.....4.5 - 4.0 bya

Five Big Start Ups: *Billions of Years Ago*

If land and sea were not present when Earth began, how did they develop? As usual, scientists are not in total agreement, but broadly speaking, they think this is what happened.

● Land was formed during Earth's early life as volcanoes arising from Earth's heated center threw up lava and rocks. As it cooled, the crust solidified and formed slabs or what are called tectonic plates that make the ocean bed and continental land masses. The rocks, corroded by acid rain and broken down by freezing, formed layers of clay and sand.

Earth's crust is still being renewed by volcanic ruffling. If it were not, it would eventually be worn smooth by the weather. And then our entire planet would be covered by an equal depth of water and we would have no mountains, no valleys, and no dry land at all. Volcanoes are our ferocious friends whose chaotic housekeeping we could not do without.

● Our oceans pose a different kind of problem. Where did so much water come from? Some scientists think it is due to condensation from the atmosphere but others think this doesn't explain the huge volume of water on Earth. They think it was deposited here by comets during Earth's earliest age. Half a comet's mass may be ice, and several large bombardments would account for most of Earth's water.

Chapter 2: Our Earth - *Born in Fire*

Chapter 3

First Life

3.9 to 1.8 Billion Years Ago

Five Big Start Ups: *Billions of Years Ago*

The Age of Microbes

That momentous event, the arrival of life on Earth, happened here almost 4 billion years ago. It started as a simple microbe and for 3 billion years - almost three-fourths of all the time Earth has been in existence – living organisms never grew larger than a single cell. Yet small as they were, some of these tiny microbes slowly, almost imperceptibly, transformed Earth's environment. They developed photosynthesis, a process that uses the sun for energy and releases oxygen as a waste product. For billions of years, they dribbled oxygen atom by atom into the atmosphere. The effect they ultimately had on the nature of life on Earth was nothing short of revolutionary.

Many of the organisms responsible for this revolution did not survive, and when oxygen finally flooded the whole world, Earth witnessed its first mass extinction. It was followed by the evolution of complex organisms and a partnership between the first tiny plants and primitive animals on which we still depend.

The pattern of thriving life followed by mass extinction which then leads to the emergence of a more complex form of life is not unique. Earth has seen at least eight mass extinctions and hundreds of smaller extinctions as species evolve, multiply, thrive, and then die out. It is not unusual that a species dies out, like these first microbes, as a result of a changing environment which they themselves bring about.

Life on Earth

3.9 billion years ago

There are no fossils from this time, so we don't know exactly how or when life appeared here. We do know that about half a billion years after Earth began, bacteria were living in the sea. It may have started there, in the mudflats around its edges, deep in Earth's damp crevices or in ocean vents where microbes still survive against all apparent odds.

These early single-celled organisms still exist in places without oxygen. We think of them as extremophiles, the ultimate survivors in bogs, in the intense heat of volcanic vents, or pools of sulphuric acid. They were suited to conditions on young Earth, and would flourish here for almost 2 billion years.

Archean Eon.....4.0 – 2.5 bya

Five Big Start Ups: *Billions of Years Ago*

more about...
are we alone?

By cosmic standards, life appeared here a short time after Earth began. We know that the only form of life that emerged and continued to evolve is the life we share. DNA analyses show that all forms of life on Earth today evolved from a single source. But there are still many unanswered questions.

● Did life evolve spontaneously on Earth or was it carried here from outer space? If it came from space, how did it get there in the first place?

● If life exists elsewhere in the Universe, does is always consist of the organic compounds of carbon, hydrogen and oxygen like all life we know on Earth? If life can exist in what we call non-organic compounds like stone, would we recognize it as a living organism?

● Does life appear only rarely, or does it appear routinely wherever matter is evolving?

The vastness and age of the universe suggests that extraterrestrial life might be common. But if it is, then as the physicist Enrico Fermi once asked, "Where is everybody?" Some centers here on Earth are sending messages into space to try to contact other intelligent life with whom we may share this Universe. So far no public, verifiable evidence has put us in contact with anyone from outer space.

Other scientists think the conditions required for life are so rare that it may not have happened anywhere else ever before and may not happen ever again.

If they are right, our being here at all is simply extraordinary. We may be the only life in all the universe. We may be that special.

Chapter 3: First Life – *The Age of Microbes*

Solar-Powered Bacteria

3.8 billion years ago

Shortly after life appeared here, a group of blue-green bacteria or "cyanos" developed Earth's first solar-powered energy system. They did it through photosynthesis trapping the energy of sunlight to break down carbon dioxide and water which they used to replenish their own energy stores – in other words, to use as food. What turned out to be absolutely revolutionary for the fate of life on Earth is that the waste produced by photosynthesis is oxygen.

Since there was an abundance of sunlight and water, photosynthesis was a stunning success and cyanos spread rampantly around the globe. For two billion years, oxygen trickled into the atmosphere. Eventually and almost single-handedly, these microbes engineered an unparalleled transformation of Earth's living conditions.

Archean Eon.....4.0 – 2.5 bya

Five Big Start Ups: *Billions of Years Ago*

more about...
photosynthesis: a revolution

The process of photosynthesis changed Earth's environment more dramatically than any other single process ever to take place here. How did it happen?

- The oxygen created as a waste product of photosynthesis gradually accumulated. First it combined with the surface iron on Earth, oxidizing and rusting it. After about a billion years, when the world was completely rusted, free oxygen began to accumulate in the atmosphere.

- After two billion years, the atmosphere had changed drastically. Life that had thrived without oxygen in Earth's early days retreated to dark corners. In the end, the sunlight world belonged to the oxygen-lovers.

- Oxygen in the form of 02 was also being transformed at the outer edges of Earth's atmosphere into 03, or ozone, the layer that protects us from the sun's ultraviolet rays.

It is difficult to exaggerate the transformation of Earth resulting from photosynthesizing bacteria and plants. Life would have proceeded on a very different path without them and if they did not continue to convert the energy of sunlight for us, we could not survive today.

Chapter 3: First Life – *The Age of Microbes*

First Supercontinent

2.5 billion years ago

We walk on it, grow corn and barley on it, even build whole cities with skyscrapers on it. Land doesn't seem to move around, and things are generally there when we go back to them. But land isn't fixed at all. Over time, Earth's relatively brittle crust has been broken by heat from below and meteors from above into 8 large and 7 smaller pieces called tectonic plates. They are floating on Earth's semi-molten mantle, jostled around like logs on water.

The history of Earth's earliest land masses is still somewhat mysterious. We do know that 2.5 billion years ago, the plates were huddled together as a supercontinent. What would become North America lay at its center. Australia and Antarctica lay on its western side, Africa on its eastern flank. Since then, the continents have bunched together and split apart again three or four times. This constant rearranging of Earth's basic floor plan is quite unsettling and continues to create some pretty drastic upheavals.

Proterozoic Eon.....2.5 bya – 542 mya

Five Big Start Ups: *Billions of Years Ago*

more about...
what happens when continents move around

Thermal currents rising from Earth's heated core and volcanic eruptions push the continents, moving them imperceptibly like great boats floating on Earth's semi-molten mantle. This process of continental drift moves the tectonic plates about as fast as fingernails grow, which might seem so negligible, at such a gentle pace, that it could barely be important.

But since they first formed, this slow tempo has moved the continents all over the globe. Africa has been under ice, America and Europe have been under the sea, fertile lands have been desert. Every continent in the world has been some place else more than once. We don't feel it or see it, but continental drift has been a critical influence on the environment and on the evolution and extinction of species.

● Continental drift causes old habitats to dry up and others to flood; it causes temperatures to change from equatorial warmth to polar cold and back again.

● When continents collide, volcanic disruption increases and continental edges, pushed up against each other, create mountains.

● In order to relieve pressure, earthquakes erupt at fault lines, that is, the points where continents rub against each other. If the heaving plates are in the ocean, they push the water into huge waves, tsunamis as tall as 100 yards. That's about the height of a 35-storey building

Continental drift continues on its inexorable way today. Since we have built cities from San Francisco to Tokyo, from Mexico City to Manila, over fault lines, the effects are often devastating. Our almost instant global communications system may make us more aware of the destructiveness of our unstable footing, but the movement is not a process we can stop.

Chapter 3: First Life – *The Age of Microbes*

The Great Oxygen Catastrophe

2.4 billion years ago

The Great Oxygen Catastrophe was a result of the bi-product all biological organisms produce in the process of expending energy. Today we often call it pollution.

At first, oxygen released through photosynthesizing bacteria was absorbed by organic matter or captured by iron. But eventually, excess oxygen began to accumulate in the atmosphere.

This oxygen eventually created pollution on such a scale that it became lethal for many bacteria. They were anaerobic, which means they were poisoned by the oxygen, just as we would be poisoned by the carbon dioxide we exhale. The atmospheric oxygen led to devastating environmental change by triggering one of the longest glaciations and possibly the greatest mass extinction Earth has ever witnessed.

As with other great extinctions yet to come, some organisms adapted to the changed environment. A few bacteria – called aerobes or oxygen-lovers – evolved respiration, a way to use oxygen to release energy. It's what occurs when we breathe. Like photosynthesis, it was a radical metabolic innovation. In fact, if it were not for the aerobes, we might all still be plants.

Proterozoic Eon.....2.5 bya – 542 mya

Five Big Start Ups: *Billions of Years Ago*

more about…
recycling pollution

The environmental change that resulted in Earth's first mass
extinction was the world-wide accumulation of oxygen – the
waste product of photosynthesizing. Some organisms adapted
and flourished by developing respiration to recycle oxygen rather
than being suffocated by it. Here is how it works:

> Oxygen combines with sugars such as acid and alcohol
> molecules, releasing energy that can be used for walking and
> talking and staying warm, and doing everything else living
> involves. The waste product of this process is carbon dioxide.
> In systems like ours, fresh oxygen and waste carbon dioxide
> are ferried by our blood cells from and to our lungs where an
> exchange is made with the atmosphere. If we cannot get
> fresh oxygen to replace the carbon dioxide we exhale, we die.

The anaerobes illustrate a rhythm that has been repeated again
and again on Earth. A species adapts and survives, it grows and
prospers, and then, sometimes after billions of years, it declines
or even becomes extinct. Sometimes this happens because the
environment changes, sometimes food runs out, and sometimes,
as is the case in this first mass extinction of anaerobes, species
decline or die because the species itself has polluted their
environments so badly with their waste products that places
where they had thrived become poisonous or produces, as in this
case, dramatic and potentially deadly environmental change.

We are facing a problem not unlike the one that led to the world's
first mass extinction. We call the waste products of our energy
consumption "pollution." There is so much of it we may be
endangering millions of species and even our own way of life. But
although we may be able to act more responsibly in relation to
our excessive waste, we cannot eliminate pollution. Merely being
alive produces waste that comes from burning energy.

Like the aerobes that developed respiration to use the excess
oxygen of bacteria, perhaps we can find ways to recycle the by-
products of our energy-exuberant lives. Bacteria have
demonstrated that recycling may be our best survival strategy.

Chapter 3: First Life – *The Age of Microbes*

49

Chapter 4

True Plants and Animals

1.7 to .5 Billion Years Ago

●Five Big Start Ups: *Billions of Years Ago*

Evolutionary Try-Outs

For more than another billion years after the Great Oxygen Catastrophe, life exploded again and again in a series of fantastic evolutionary try-outs in the oceans of the Earth.

It is impossible to date with precision when various life forms first appeared. Some organisms seem to have appeared quite suddenly, but whether this is actually what happened or whether their forebears have simply disappeared without a trace, lost to the waters in which they evolved and died, it is impossible to tell.

Although some of the organisms that flourished during this time have disappeared into the mists of time, some survived for hundreds of millions of years before succumbing to environmental changes. Others have evolved into life forms that are still with us today. And some amazing plants and animals from this time are still living in Earth's seas today, almost unchanged for perhaps as long as a billion years.

Cells with a Nucleus

1.7 billion years ago

What happened after the great oxygen poisoning is one of the great turning points in Earth history. Single-celled bacteria developed a central nucleus and began to incorporate other bacteria. But instead of being digested, the captured bacteria were kept intact so they were able to contribute their special capacities to the enlarged cell.

The first plants evolved by incorporating blue-green bacteria which gave them photosynthesizing capacity. Tiny pre-animals evolved by incorporating aerobic bacteria, giving them the mechanism for respiration used by all animals.

It was the beginning of a perfect partnership on which we still depend today. Plants take in carbon dioxide and water and expel oxygen while animals do just the opposite, taking in oxygen and expelling carbon dioxide and water.

Proterozoic Eon.....2.5 bya – 542 mya

Five Big Start Ups: *Billions of Years Ago*

more about…
a long-term relationship

Although they were still only a single cell, the new organisms began the basic mechanism of sex. They are called "eukaryotes" – pronounced "you carry oats." It's not the name of every Tom, Dick and Harry, but every one of our own body's cells is a eukaryote, so perhaps it's worth passing on the name. It derives from the Greek meaning "with a nucleus," which distinguishes it from the primitive prokaryotes or cells "without a nucleus." The eukaryote nucleus contains genetic material and reproduces by mixing sets of chromosomes from two different parent cells.

Eukaryotes were very small, but they began the ecosystem we still rely on. They evolved into three interdependent types:

● Algae were plants that kept sunlight-using bacteria alive after incorporating them. They collected as colonies in shapes like plates, balls or stalks and were so successful that even today 90% of the cells on Earth that use photosynthesis are algae.

● Protozoa were microscopic pre-animals. All animals, as opposed to plants, use respiration to create energy so they cannot make their own food. Without plants as their ultimate food source, animals would starve. And that includes us.

● Fungi evolved by incorporating the capacity of the early oxygen-hating bacteria and are among the world's great recyclers. They break things down into their original components and are usually responsible for the decomposition when things rot. We might think of them as spoiling things, but the truth is that we need them to clean up after us.

Chapter 4: Plants and Animals – *Evolutionary Try-Outs*

Multicellular Life

900 million years ago

A little less than a billion years or about 900 million years ago, an extraordinary evolution occurred. Multicellular organisms, true animals and plants larger than a single cell, appeared. Why it happened, apparently quite suddenly, after almost 3 billion years during which all organisms were a single cell, is not clear. But happen it did.

The earliest of these new organisms were soft-bodied and have left behind incomplete fossils that only hint at their structure. 700 million years ago, animals with shells had evolved. Yet by 545 million years ago, much of this new life had disappeared, probably driven to extinction by severe climate gyrations and fierce competition.

Proterozoic Eon.....2.5 bya – 542 mya

Five Big Start Ups: *Billions of Years Ago*

more about…
multicellular experiments

Some amazing fossils from this time were first found beached in the Ediacaran Hills in Southern Australia. They all lived in the sea, did not have a backbone, and experimented extravagantly with new ways to sustain life. They were hard and soft, stationery and free-floating, solid and transparent, some had incipient heads, others were equipped with defensive armour.

● Fantastic Bewilderments
Some fossils are so strange they still baffle scientists. There were blobs and globules, some exquisitely patterned in round or worm-like shapes. An organism that looks like a cross between a cauliflower and a bed quilt seems unrelated to any other creature we know. Some seem to have reached an evolutionary dead end, to have simply failed to thrive.

● Familiar Survivors
A few animals from this time begot an impressive lineage. Sponges still root to the seabed like plants and get nutrients by filtering water drawn in by beating tiny hairs. Jellyfish have no brain but do have muscles and nerves. Molluscs developed a defensive shell and a gill for respiration. With jointed legs and bodies, early arthropods are among the world's great survivors, and today outnumber all other types of living animals combined.

● The Fabulous Flatworm
Most of us don't place flatworms high on our list of great evolutionary achievements. But the evolution of most successful animals, including us, seems to lead back to them. Flatworms developed bilateral symmetry, bodies with a right and left mirror image. If we had descended from the symmetrical jellyfish instead, we might all be round balls without a backbone.

Chapter 4: Plants and Animals – *Evolutionary Try-Outs*

The Cambrian Explosion

542 million years ago

The extinctions of the first multicellular organisms were followed by a truly astonishing period called the Cambrian Explosion. Millions of new species, again apparently quite suddenly, evolved in the sea. Some were bizarre, even phantasmagorical; others are of critical ancestral interest to us. All the major animal groups we know appeared. The last group or phylum to develop was the chordate, animals with a spine. It's the phylum we belong to.

Cambrian Period.....542 - 488 mya

Five Big Start Ups: *Billions of Years Ago*

more about…
the great cambrian explosion

When fossils dating back more than 500 million years were found in British Columbia, some paleontologists thought we had stumbled on the very dawn of life. There were creatures with shells, tentacles, spikes, scales, antennae, stalks, and eyes.

Two species illustrate this prolific, marvelous time.

- Trilobites: One third of all the fossils from this period are trilobites. They belonged to the arthropods, ancestors of today's insects and crustaceans. They ranged from microscopic smallness to a foot, and developed eyes with thousands of lenses that made them exceptionally clear-sighted. They survived for 300 million years.

- Pikaia: Like the flatworm, the pikaia may not be the first relative we would introduce to the family. It looked like a two- inch worm that could swim, but it was a chordate, the earliest animal we know that had a rod for stiffening its body. This eventually evolved into the backbone of vertebrates like fish, amphibians, and mammals, which makes them our own primordial ancestors. Their nerve cord developed a front-end swelling that eventually became a brain with senses of smell, touch, and sight.

There are some significant advantages in being a vertebrate. An internal skeleton can grow with the animal so it does not need to be replaced and a bone case around the head protects the brain and critical senses. It might not look like much, but the Pikaia is a relative we can be proud of.

Chapter 4: Plants and Animals – *Evolutionary Try-Outs*

Chapter 5

First Land Life:

470 to 250 Million Years Ago

Five Big Start Ups: *Billions of Years Ago*

Strange and Dangerous

By about half a billion years ago, atmospheric oxygen had reached today's levels and an ozone shield had developed which protected land creatures from damaging sun rays. Plants were the first to move out of the sea and onto land. Fish followed, the first vertebrates to survive outside the sea. The move to land life was no less epic than man's exploration of outer space.

It took hundreds of millions of years of evolutionary change before this first eon of terrestrial adaptations were complete. The time was punctuated with several ice ages each lasting millions of years and at least three mass extinctions. Despite stunning innovations that gave new species considerable mastery over the land, the era ended with an extinction event so grim that it came close to annihilating most of life more complex than microbes.

Terrestrial Plants

470 million years ago

Some forms of bacterial life may have begun to survive on sandy soils as long as a billion years ago. But until recently – recently anyway in terms of life on Earth - land on Earth was almost barren as the surface of the moon or Mars we see today. Then, a little less than a half billion years ago, a successful voyage on to land began, probably led by a cheeky ancestor of the inventive lichen. A fungus with a resident blue-green bacteria, lichen even today can be counted on to show up first on desolate rock swept clear of all visible means of sustenance. The ultimate impact of the conquest of land by plant life was immense.

In the sea, life continued in exuberant experimentation. We who evolved much later to walk upright may find it particularly noteworthy that this is when fish with a backbone first appeared in Earth's waters as true vertebrates. When plants had established a stronghold and could provide food and shelter, many of these animals eventually followed them onto dry land.

Ordovician Period.....488 - 444 mya

Five Big Start Ups: *Billions of Years Ago*

more about...
problems for land immigrants

Land at first was as alien to sea creatures as outer space is to us today and these first ancient steps onto land were rather like our astronauts' first steps on the moon. Terrestrial immigration was strewn with many failures.

A mild climate that remained stable for fifty million years helped plants and then animals devise an impressive variety of ingenious strategies to meet the unprecedented demands of land life. Still, it took 200 million years to work out enduring ways to

● avoid drying out in the sun.

● maintain bodily structures that were no longer supported by the buoyancy of water.

● extract oxygen from air instead of from water.

● adapt to the violence of land storms and the greater temperature fluctuations of air compared to water.

● move around on land instead of water.

● develop methods of reproduction that were successful outside a watery environment.

Chapter 5: First Land Life - *Strange and Dangerous*

Mass Extinction of Plants and Animals

445 million years ago

The high spirits of the previous fifty million years came to a dramatic halt between 450 and 440 million years ago. A supercontinent called Gondwana - composed of what are today South America, Africa, Antarctica, India, Australia, and even for a while Florida and part of south western Europe - had been moving toward the South Pole. When it settled there, huge glaciers formed and sea levels dropped, causing the first mass extinction of our present age. A second wave of extinctions may have followed a million years later when glaciers retreated and sea levels rose rapidly again.

85% of living species died. The losses were enormous, and may be the second most devastating of all mass extinctions ever to take place on Earth.

Ordovician Period.....488 - 444 mya

Five Big Start Ups: *Billions of Years Ago*

more about…
mass extinctions

Extinctions are not an unusual occurrence. In the history of Earth, the extinction of species, like the death of individuals, is the norm. The average species survives about four million years, and 99.99% of all species that have ever lived on Earth are now extinct.

In the last 550 million years there have been 54 major extinction events we know about. Five of these have been so devastating that multicellular life almost disappeared.

- 445 million years ago 85% of species perished.

- 365 million years ago 70-85% of all species perished. It was worse for marine than land life.

- 250 million years ago the most catastrophic extinction Earth has ever endured annihilated 95% of all species.

- 206 million years ago a mass extinction opened the way for the dinosaurs.

- 65 million years ago another mass extinction ended the reign of the dinosaurs.

At present, a major extinction may be going on, though scientists disagree if extinctions are as catastrophic as we fear. The conundrum is that human activity may actually be enhancing the evolution of some species while destroying others. Historically, global warming has supported the evolution of new species. But hundreds of advanced and newly-evolved mammals including primates that will not be replaced easily are presently under serious threat.

Chapter 5: First Land Life - *Strange and Dangerous*

Land Animals

440 million years ago

In Canada today, there are tiny tracks made by an intrepid little explorer rather like a mite, with jointed legs and a determination to find a better life. It was one of the first animals we know about to leave its watery home to explore the land beyond and begin animals' mass migration out of the sea.

After the earlier glacial conditions, the climate became warmer again for another 30 million years and insects, silverfish, centipedes and spiders transferred to a land home with enthusiasm. Arthropods like cockroaches, dragonflies and scorpions solved the problem of drying out with a stiff skin and waxy coat, and developed mouth parts strong enough to munch on a burgeoning feast of plants. Marine vertebrates multiplied too, and the first coral reefs appeared.

It was a wonderful time that lasted for millions and millions of years.

Silurian Period.....444 - 416 mya

Five Big Start Ups: *Billions of Years Ago*

During this period, a fish called a placoderm evolved. It was encased in armor plating protecting it against predators like the vicious 7-foot sea scorpions prowling the seas. The first fish vertebrates also evolved a defence more useful to us than armor plates. It is an immune system unique to vertebrates, a marvellous system, with a defensive coordination that still outstrips even the most advanced techniques of modern medical science.

- There are millions of specialized cells called lymphocytes in the body, each with the ability to fight a different potential invader. When a lymphocyte meets a foreign cell which is its counterpart, it produces an antibody, a protein that attaches itself to the invading cell.

- The lymphocyte then multiplies. Some attack and kill the invader, while others release more antibodies for defensive action throughout the body.

- Cells called phagocytes provide a mopping up action by destroying any cells with an antibody sticking to them.

Once an invading cell has been attacked, the immune system remains primed, sometimes for life. That's why we - fortunate descendants in possession of the immune system bequeathed to us by the fish - don't get chicken pox or measles or mumps more than once. Vaccines make use of this immune system by arming it ahead of time against invasion.

It's such a fantastic defence that one almost wonders how any of us ever gets sick. But alas, viruses and bacteria are also amazingly inventive, and continue to evolve ways round it.

Chapter 5: First Land Life - *Strange and Dangerous*

Trees and Mountains

420 million years ago

During this lush period, vast tracts of Earth's barren land began to turn green for the first time. Ferns grew 20 feet tall, wood-making trees 60 feet, a luxuriant spread that would provide oxygen and food for animals. Some of the oldest mountains in the world today were also forming. The mountains of Scandinavia, the Scottish Highlands, and the Appalachians are on three different land masses now, but they were formed during this period as a single mountain chain when the continents, made up of what later became Europe and North America, ran into each other.

Silurian Period.....444 - 416 mya

Five Big Start Ups: *Billions of Years Ago*

more about...
how the plants did it

Over time, plants devised strategies to solve the challenges of a life on dry land.

● 470 million years ago, they developed vascular tubes supported by fibers which could transport and preserve moisture and enabled them to grow much taller without flopping over. This solution was such a success that plants everywhere still use it.

● Although plants such as ferns grew very tall, they could not spread far onto dry land because they needed water to reproduce. Their spores floated through the air and sprouted after landing. Then it shed both egg and sperm into water where they united to produce a new plant.

● 300 million years ago, seed-bearing plants developed a way of reproducing without surface water by using the wind to spread sperm-bearing pollen to the egg. Once the egg was fertilized, the embryo was enclosed with a food supply in a protective cover and dropped to the ground as a seed. When the weather conditions were right, the seed started to root and grow.

These gymnosperms, as they are called, spread across the land. The Earth began to turn green for the first time.

Chapter 5: First Land Life - *Strange and Dangerous*

Fish Adapt to Land

370 million years ago

Today we think walking on water is miraculous, but 370 million years ago, the challenge was walking on land. A land home was enticing, though, as sharks and other predators stalked with deadly speed, making the oceans dangerous places. Whether they were hardy, desperate, or courageous, some fish crawled out of the water. They evolved as amphibians, the first four-legged animals on land. Like frogs today, they lived as adults on land, and returned to water to reproduce.

Devonian Period.....416 - 359 mya

Five Big Start Ups: *Billions of Years Ago*

We owe a lot to fish. They gave us our backbones, our immune systems, and our four limbs. They are our ancestors who first immigrated to land, where eventually we learned to use fire and develop a technology not possible to full time inhabitants of water. Fish that successfully transferred from the sea obviously had to make some adjustments, a transformation that took millions of years.

Evolutionary changes helped them adapt to land:

● Their swim bladders became lungs which they used to extract oxygen from the air instead of from water.

● Their bone cage became more robust and flexible, and muscles developed to support their weight on land.

● Bones originally supporting the fins nearest the head evolved into limbs with wrist and elbow joints and digits.

● Back fins developed into legs and in one branch, gill plates grew into wings.

Because they developed four limbs, the first "land fish" are called tetrapods, lizard-like creatures with snouts and tails. Four legs eventually became, among other things, two legs and two arms. We think, like the flatworm and pikaia, they must be among our distant ancestors. But there is a small problem: all four-limbed animals today, including birds, humans, cats, dogs, whales, and fish, have five digits at the end of each limb. The only tetrapod fossils from this time found so far had eight digits.

Chapter 5: First Land Life - *Strange and Dangerous*

A Second Mass Extinction

365 million years ago

After 90 million years of mostly warm weather, the climate cooled. The environmental shifts destroyed many habitats and 70-85% of all species may have perished in the largest mass extinction of marine invertebrates of all time. Reef builders and animals living in shallow waters were hit hardest, while species newly immigrated to land were less devastated. The extinctions continued for at least half a million, possibly even for an excruciating 15 million years.

Devonian Period.....416 - 359 mya

Five Big Start Ups: *Billions of Years Ago*

more about...
causes of major extinctions

Some extinction events have lasted for millions of years; others may have occurred in days. Surprisingly, we have only educated guesses about what caused most of them. They come from...

Outer Space: • Meteor and comet Impacts have left 30 craters wider than six miles on Earth. Each one must have been catastrophic. • If solar flares ever licked out in a frenzy of unusual ferocity, they would leave little evidence of their murderous visitation. • As it traverses the galaxy every 240 million years, the solar system moves through a rain of potentially deadly radiation.

Seriously Bad Weather: • Tectonic shifts relentlessly create and destroy shallow coastal waters where so much life thrives. They also cause earthquakes, profoundly affect the climate, create mountains and deserts, and influence the onsets and endings of ice ages. • Tsunamis, gigantic waves propelled by convulsions of the ocean floor, surge over land, drowning almost everything in their wake. • Ice ages occur about every 250 million years and last for millions of years. Water levels drop and lands disappear under ice. When climates warm, flooding destroys coastal habitats again.

Supervolcanoes: • Giant eruptions bursting from deep below the Earth's crust create massive lava flows and throw up debris so dense it reduces our sunlight for years.• Methane gas is always leaking from the ocean floor, and occasionally it explodes. A giant explosion could explain at least one of the great mass extinctions. • Earth's periodic atmospheric changes, especially of oxygen levels, create and destroy habitats.

Competition: • Species compete with each other to the death. Through war, hunting, epidemics, and wholesale environmental change, living organisms from bacteria to humans cause extinctions.

Chapter 5: First Land Life - *Strange and Dangerous*

Age of Coal

360 million years ago

During this time, coal, the fossil fuel that has contributed so much to our modern world, was formed. Despite the extinctions, the greening of the planet continued for another 75 million years. Huge plants and tropical forests grew trees reaching 160 feet. Supported by this explosion of photosynthesis, oxygen levels were elevated to 35%, 15% higher even than today. It helped a lot of plants grow very fast and very big. The big is what ultimately made so much coal.

Like plants, animals also swelled to nightmare proportions: spiders were 20 inches, millipedes six feet, scorpions three feet. Dragonflies had wing spans of more than two feet.

Carboniferous Period.....359 - 299 mya

Five Big Start Ups: *Billions of Years Ago*

more about…
our coal reserves

This is the age that gave us the coal for the Industrial Revolution a century and a half ago. It's the age that still powers many of our electricity plants, lights our campfires and melts our marshmallows. It's a fossil fuel and so, unfortunately, it is also responsible for a lot of our pollution.

How did it happen? Gigantic plants living in today's Asia, Britain, northern Europe and North America burgeoned in swamps, and sank beneath the water when they died. Instead of rotting, millions of years of heat and pressure turned them into coal. A lot of trees and plants lived and died for us in those days.

That, at least, is the received wisdom at present. There is another theory that fossil fuels formed deep in Earth where extremophile bacteria were living shortly after the planet formed. Some people think it's a whacky idea but that doesn't make it wrong. As Einstein once said "If at first an idea is not absurd, then there is no hope for it."

Chapter 5: First Land Life - *Strange and Dangerous*

Eggs with Shells

300 million years ago

We boil them, scramble them, decorate and hide them, we even use eggs to shampoo our hair, but we don't usually think of the egg as an evolutionary masterpiece. That is what an egg encased in a shell was, though, an ingenious method of reproduction that worked because it protected and nourished newly-conceived young on land. Birds today are the most well-known egg-layers, their method of reproduction inherited from their reptile ancestors.

Carboniferous Period.....359 - 299 mya

Five Big Start Ups: *Billions of Years Ago*

more about…
the amniotic egg

The egg inside a shell is called an amniotic egg. It was not, strictly speaking, the only method by which terrestrial vertebrates could reproduce 300 million years ago, but it was by far the most effective. The male injects the sperm directly into the female's body where it fertilizes the egg. The egg then develops a shell that protects the embryo after it is laid until it is hatched.

Like the plant seed, the egg provides for the developing offspring. Inside its protective walls, the yoke supplies nutrition, the amnion provides water and is a cushion against shock and temperature change, and waste carbon dioxide and uric acid is collected in a sac called the allantois.

Why was the amniotic egg such an evolutionary advance? Besides freeing the parents from the need for surface water to reproduce, it permitted the newborn to grow bigger before it had to fend for itself. This is a huge initial advantage, increasing the survival rate of offspring, and allowing them to evolve into more complex adults.

Chapter 5: First Land Life - *Strange and Dangerous*

Another Ice Age

290 million years ago

The age of coal with its luxuriant vegetation was terminated by another long ice age. It lasted for **42** million years in all, but the climate depended on the vagaries of continental drift. Land masses were colliding and by **250** million years ago, one great land called Pangaea, "All Land," stretched from the north to the south poles. The south pole region was buried under ice for millions of years. The central section dried out into a vast desert. The north, damp and humid, was where many land animals and plants survived.

Permian Period.....299 - 250 mya

Five Big Start Ups: *Billions of Years Ago*

more about...
ice ages

Ice ages occur regularly. In all, they last millions of years, and alternate between warmer periods called <u>interglacials</u> when only mountain tops and polar regions are permanently covered with ice, and severe periods of cold called <u>glacials</u>. It is confusing, because the glacials are also sometimes called "ice ages." Glacials usually last about 100,000 years and intervening interglacials last an average of 10,000 years.

Ice ages sometimes start when continents drift over the poles. The cold glacial and warm interglacial fluctuations then typically occur in overlapping cycles of 100,000, 41,000 and 22,000 years and are related to Earth's position in space. Like volcanoes, ice ages are important because they ruffle Earth's surface, assuring that not all our land is covered by water. They also disrupt environments and affect the course of evolution.

In the last billion years there have been three ice ages when the polar caps have been covered with the thick layers of ice with which we are familiar. The first ice age 800 million years ago lasted 200 million years. The second started about 300 million years ago and lasted about 40 million years. The last ice age began 10 million years ago and has not yet ended. Right now Earth is in an interglacial period that began more than 12,000 years ago.

Chapter 5: First Land Life - *Strange and Dangerous*

The Great Dying

250 million years ago

About a quarter of a billion years ago, Earth endured the most catastrophic extinction in history. 96% of all living species may have perished, 75% of all vertebrates. The trilobites, prolific survivors for close to 300 million years, all died. Living things came close to being knocked back a billion years when Earth was colonized only by microbes. It is called the Permian Extinction. Never before or since has the planet witnessed death like this.

Permian Period.....299 - 250 mya

Five Big Start Ups: *Billions of Years Ago*

more about…
did global warming cause the mass extinctions?

What caused this most cataclysmic of all extinctions? There are many possibilities, and there may have been more than one cause.

● Scientists have discovered an underwater crater off the coast of Australia that could be the landing site of the "sailback killer," an asteroid that crashed into Earth from outer space 250 million years ago. The impact of the asteroid would have been followed by massive volcanic eruptions, covering the skies with dense cloud for years, and instigating massive fires and giant tsunamis racing for miles onto land.

● It is possible that the Australian impact site didn't come from outer space but from a supervolcano of almost indescribable power, a gigantic gun shot shooting up from the bowels of the Earth. It would have ejected huge chunks of rock into space, triggered massive waves and fires, and left much the same evidence of its visitation as a meteor strike. Most devastating of all, huge volumes of greenhouse gases would have heated up the world in a suffocating embrace of global warming. Too much heat and too little oxygen would have killed almost everything living on land and sea.

Did all the life that perished in the Great Dying die because of an onslaught from outer space? Or from an explosion propelled from the inferno at Earth's heart? Were there other catastrophes? We don't know.

Chapter 5: First Land Life - *Strange and Dangerous*

Chunk II

The Dinosaurs

251 - 65 Million Years Ago

Geologists call the time between 250 and 65 million years ago the Mesozoic Era – the age of Middle Life. It is also more famously known as the age of the Dinosaurs. But many other things of great importance to us today also began during this era.

Although dinosaurs and mammals both began to develop as this era began, it was the dinosaurs who eventually dominated the world for more than 140 million years. This era was also a time when the evolution of birds, insects, fruits and vegetables would make it possible for mammals to take the dominant place when the dinosaurs were eventually felled.

6. 251 million years ago: The first period of Middle Life is called the Triassic period, when land was populated by <u>reptiles</u> which had survived the earlier extinction that had brought an end to 90% of Earth's most ancient life, and by the first <u>dinosaurs and mammals</u>.

7. 200 million years ago: The <u>dinosaurs dominated Earth</u> by growing faster and bigger than any other living thing. During this Jurassic Period they were almost unchallenged.

8. 65 million years ago: The <u>dinosaurs' extinction</u> was the result of another major extinction of life on Earth, an event so massive that it led to a whole new era in the history of life.

Chapter 6

Reptiles, Dinosaurs and Early Mammals

251 - 200 Million Years Ago

The Dinosaurs: *Millions of Years Ago*

The Big Three

The first stage of what is called Earth's "Middle Life" was the Triassic Period. It is something of a misnomer because it was given that name by an archaeologist to reflect three layers of rock in a part of Germany not found elsewhere in the world. There is, though, a Mega-Triad, a Big Three, that tower over this period. They are reptiles, the first dinosaurs, and early mammals.

During this time, the land masses were bunched together in a great continent called Pangea that sprawled across the equator. For 40 million years the climate was warm and accommodating. Reptiles above all multiplied and flourished and early mammals evolved.

Probably the most well-known event of this period is the appearance of the first dinosaurs. Their oldest fossils found thus far are in what is now Argentina and Brazil and are about 230 million years old. Among this group was a small meat-eating dinosaur called Eoraptor. Its skeleton suggests it might not be the most primitive of dinosaurs and archaeologists surmise that dinosaurs must have appeared somewhat earlier than the oldest fossil yet discovered.

Ancestral Reptiles

250 million years ago

They aren't cuddly and save for the odd turtle, we don't usually count them among our favorite pets. But reptiles are among the greatest and most prolific survivors on Earth. They first emerged in the sea about 315 million years ago, moved onto land when the opportunity arose, and lived through the Great Dying. They are still with us, so in one form or another, they have been around a long time. Despite their small size relative to many dinosaurs, they have far outlasted them.

Because they themselves were sometimes both huge and ferocious, some reptiles are confused in the popular imagination with dinosaurs. But although reptiles gave rise to dinosaurs, they also remained one of their few potent enemies during the dinosaurs' long supremacy.

One of the most notable reptiles of this period is a Pelycosaur reptile called the Dimetrodon, an aggressive creature that looked a little like (but wasn't) a dinosaur with a sail on its back. It was an early ancestor of the mammals, and one of the first land animals that could prey on animals its own size. The "Sailback" did not survive, and the lineage that eventually evolved into mammals and ultimately gave birth to *Homo sapiens* stayed very small for the next 200 million years. Yet, perhaps for better or worse, the spirit of the predator with a sail on its back lives on.

Triassic Period.....250 – 200 mya

The Dinosaurs: *Millions of Years Ago*

Reptiles are categorized by differences in the holes in their skulls, which might not be very interesting, except that they ultimately evolved into four different megadynasties, as it were, and we belong to one of them.

- <u>Anapsids</u> are reptiles without any skull holes. Turtles and tortoises may be the only anapsid survivors today.

- <u>Euryapsids</u> had one skull hole. They are now all extinct but during the Mesozoic Age, aquatic reptiles called Plesiosaurs were prolific euryapsids. Their fossils have been found around the world.

- <u>Diapsids</u> were reptiles with two skull holes. Most famously they include a group called Archosaurs, "the Ruling Reptiles," often huge animals with teeth firmly set in their jaw sockets, a characteristic which gave them an obvious advantage. Although often mistaken for dinosaurs themselves because of their power and size, Archosaurs were the direct ancestors of the dinosaurs. They were also the forebears of crocodiles, birds, and the Pterosaur, a huge flying reptile of this age often mistaken today for a bird. These flying retiles had no feathers but some seem to have had fur which would have contributed to their being one of the most accomplished flying machines ever seen among living animals.

- <u>Synapsids</u> were reptiles with a pair of holes on each side and one hole lower down in the skull which ultimately supported the ability to eat and breathe simultaneously. Synapsids evolved, among other things that are now extinct, into proto-mammals called Therapsids, which evolved into full-fledged mammals which eventually included *Homo sapiens*.

Chapter 6: Reptiles, Dinosaurs, Early Mammals –
The Big Three

Early Mammals

200 million years ago

After the Great Dying, a class of warm-blooded animals evolved in whom we have a personal interest. Early mammals had hair and differentiated teeth, laid eggs containing partially developed embryos and produced milk for their young. At the time we would not have picked them out as the animals most likely to dominate and transform Earth in 250 million years or so, however. The first mammals were small shrew-like animals who survived in inconspicuous niches away from the powerful and prolific reptiles.

For almost 40 million years, the world was their benevolent home. But then, 208 million years ago, another mass extinction occurred, possibly triggered by an asteroid. It probably delayed the ascendancy of the early mammals, making way, instead, for the reign of the dinosaurs which was about to begin.

Triassic Period.....251 – 200 mya

The Dinosaurs: *Millions of Years Ago*

Evolution is sometimes called the theory of the survival of the fittest. That is not quite right. Species that survive aren't the "fittest" so much as those that "best fit" a new environment that may be hotter, colder, drier, or wetter than before. Survivors are often more complex but not always. The most adaptable may be smaller, simpler, tougher, stronger, or more aggressive, not smarter, bigger, more creative, or more intelligent.

After the Permian extinction, mammals had what looked like some significant evolutionary advantages. They are warm-blooded and have arteries that carry oxygen-rich blood to the cells, and veins that carry the oxygen-depleted blood back to the heart and lungs. Keeping the oxygen-enriched blood separate supplies more oxygen to the whole body and is especially valuable to the mammal's superior brain.

But the great day of the mammal – and so ours – did not arrive until after the dinosaurs' period in office. During all that long time, mammals grew no larger than a small cat, until another extinction event displaced the dinosaurs' supremacy.

Chapter 6: Reptiles, Dinosaurs, Early Mammals –
The Big Three

Chapter 7

Dinosaurs Reign Supreme

208 – 145 Million Years Ago

The Dinosaurs: *Millions of Years Ago*

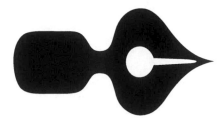

Jurassic Park

It was dinosaurs, not brainier mammals, who gained ascendancy after the extinction that ushered in the Jurassic Period. They survived for some 170 million years during which time they ruled the world in an unprecedented reign of 140 million years.

As the climate warmed, insects and plants co-evolved, creating a mutually beneficial ecosystem that has become a vital food source for the whole world today.

The Dinosaurs

200 million years ago

Dinosaurs, the "Terrible Lizards," lived for at least 180 million years and subjugated the world for more than 140 million. At first they were small and agile meat-eaters, but over the millions of years of their domination, they came in all sizes.

They had few predators who were their match, except for some of the largest reptiles, crocodiles that survive to this day

They weren't beautiful, and their brains were small, but they were strong and often big, and they survived a very long time. Dinosaur fossils today are found throughout the world, which suggests that in their time, dinosaurs walked on every land mass on Earth.

Jurassic Period.....200 – 145 mya

The Dinosaurs: *Millions of Years Ago*

more about…
dinosaurs

By the criteria of survival, dinosaurs were very successful. What's more, during most of the time they lived, they strode around the globe, the most important animals in existence. What were these incredible creatures?

● All the dinosaurs had long tails and limbs, but apart from that, their appearances varied. Some had feathers, some were protected with scales. They were as short as a few feet or as tall as 110 feet and as heavy as 30 tons. Some had long necks and could rear up higher than a seven-storey house. Some were squat precursors of birds that looked a little like terrifying turkeys. Others resembled ostriches with clawed hands and bony tails.

● Dinosaurs were grazers, scavengers and hunters. They ate insects, plants, animals, and sometimes other dinosaurs. With their long tails and short legs, they may have been prolific swimmers and used the sea as a significant source of food. Some of them seemed to have lived in families and spent years caring for their young.

● Dinosaurs, like snakes today, may have been cold-blooded, and able to keep warm only by basking in the sun. If so, they fed during daylight, leaving the dark to the small scurrying mammals who were equipped to deal with night-time temperatures. Some dinosaurs may also have been warm-blooded or something in between.

Chapter 7: Dinosaurs Reign Supreme – *The World Park*

Birds and Insect Colonies

180 million years ago

True birds began to evolve at this time, and today may be the only remnants of the once mighty dinosaurs.

Insects, which already had been around for quite a while, were now diversifying. 120,000 species of flies, mosquitoes, gnats, midges, and no-see-ums, along with moths, butterflies, bees, ants, wasps, beetles, roaches, termites, grasshoppers, and aphids all flourished. Insects are pretty little, so we don't think about them a lot. But as we shall see, insects and plants were about to develop a partnership that even today is more essential to our welfare than most of us appreciate.

Jurassic Period.....200 – 145 mya

The Dinosaurs: *Millions of Years Ago*

more about...
what makes insects interesting

We step on them, exterminate them, spray them and erect nets to keep them away. Except perhaps for a friendly bee whose honey we gather, we generally think of insects as pests.

Yet insects are really quite amazing. A whole moving, sensing, living, reproducing system operates in the tiny space of their small bodies. There are probably several million insect species, most of which live on land rather than water. Some insects are almost unaccountably cooperative. Termites and then ants and bees formed the first communal colonies dedicated entirely to the common good where members engage in specialized tasks, gathering food or defending the fort, while a special few reproduce.

The Greeks and Romans admired the bee hive as a model of industrious cooperation. With the invention of the microscope, the discovery that the "king bee" was a queen was so scandalous that in 1744, an English clergyman insisted she must be a virgin who could not be so debased as to have sex with more than one male.

Chapter 7: Dinosaurs Reign Supreme – *The World Park*

Flowers and Fruit

150 million years ago

Flowering plants evolved quite recently, just in time for warm-blooded mammals like us who need high-energy food to maintain our body temperatures and keep our brains working. Flowering plants give us our fruit, our vegetables, and grains, our breakfast cereal, our bread and hamburger buns, our pastas and rice pilaf. Beans, nuts, herbs and spices come from them and so do beer, wine, cola, coffee and tea. Our chocolate is sourced from flowering plants, as is everything we make from cotton or linen or hemp. They give us medicines like aspirin and morphine, and recreational substances like tobacco, alcohol, marijuana and cocaine.

Flowers aren't just beautiful. Our lives depend on them. No wonder they are growing in gardens and fields all over the world.

Jurassic Period.....200 – 145 mya

The Dinosaurs: *Millions of Years Ago*

more about…
how the garden grows

In a marvellous serendipity, flowering plants evolved hand-in-hand with pollinating insects like bees, ants and butterflies in a mutual interdependence that has eventually come to include us.

Flowering plants attract insects with scent and color. Some colors can be detected only by insects who can respond to ultraviolet rays we humans can't even see with the naked eye. Inside the flower is a sugary food called nectar. While they are extracting the nectar, insects also pick up pollen from the flower, which they carry to the next flower in their search for more nectar.

The pollen has two different sperm. One fertilizes the egg, which creates a seed. The other fertilizes the flower which grows into the fruit that nourishes the seed during the time it is sprouting.

Flowering plants can make fruit in weeks while it can take over a year to store enough food in a seed alone to sustain a new plant until it puts down roots of its own. So flowering plants reproduce faster and are much more prolific than plants that depend only on seed.

Chapter 7: Dinosaurs Reign Supreme – *The World Park*

Chapter 8

Dinosaurs' Extinction

145 – 65.5 million years ago

The Dinosaurs: *Millions of Years Ago*

Chalky Crash

The Cretaceous Period is, literally, the period of chalk. During this time, the supercontinent of the Triassic period separated into the continents we know today and many of the buried treasures we are digging up to support our modern life styles were laid down during this period.

At the same time, while the dinosaurs strode around the world in seemingly unassailable dominance, small shrew-like mammals evolved the placenta, creating a reproductive method that greatly enhanced the potential survival of newly-born offspring. When the dinosaurs were finally dethroned, mammals were ready to fill the vacancies.

Oil and Chalk

145 million years ago

As Pangaea, the great land mass that had gathered together about 250 million years ago, broke up, the outlines of the continents we know today began to take shape. Volcanoes are more frequent when the tectonic plates are colliding and the unsettlement of this time contributed to more of the fossil deposits upon which our modern world now depends. Crude oil was formed when plankton living in the warm waters died and decayed, and chalk deposits are the remains of marine animals.

Cretaceous Period.....145 – 65 mya

The Dinosaurs: *Millions of Years Ago*

Squirrels bury nuts, dogs hide their bones, pirates have buried their ill-gotten loot. But the greatest of our treasures were buried by Mother Earth. Digging them out has often been dangerous, hard, and dirty work. But the rewards have been so beneficial and so lucrative that men, women and children have been quarrying for at least 30,000 years. What are we digging up?

● We extract fossil fuel and deposits formed after organisms died – coal and peat are from plants and trees, oil from plankton, chalk and limestone from shells of marine animals.

● Sometimes we dig up rocks – marble, sandstone, gravel, granite, aggregates and clay.

● We mine for minerals and metals – gold, silver, copper, tin, lead and iron ore, the iron oxidized by photosynthesizing bacteria more than two billion years ago.

● Gems and crystals - diamonds, turquoise, obsidian, emeralds, opals, salts and graphite - are mined for their beauty and for industry.

It's amazing what treasures are beneath our feet.

Chapter 8: Dinosaurs' Extinction – *Chalky Crash*

Placental Mammals

125 million years ago

Although the rest of us might be forgiven if we initially lack enthusiasm for the very old remains of a five-inch mouse, scientists were overjoyed when they found the fossil of an animal that had lived in China 125 million years ago. It was a mother mouse, the earliest fossil ever unearthed of a placental mammal. These are mammals that do not lay eggs but give birth to live young,

Other fossils found recently in China reveal that by this time other mammals were no longer small shrew-like herbivores but aggressive carnivores about the size of a dog. We know now that mammals were living and possibly competing with the dinosaurs for 60 million years. When the dinosaurs died, mammals took their places.

Today there are some 4,000 species of placental animals. They include rodents and bats, whales and elephants, armadillos, dogs, horses, cats, and all the primates. And of course, us humans.

Cretaceous Period.....145 – 65 mya

The Dinosaurs: *Millions of Years Ago*

more about…
the dawn mother

The proto-mouse found in China is called the "dawn mother." It was christened with this momentous name because the development of a placenta where the fetus can grow inside the mother's womb is a critical indicator of advanced evolutionary development. Why?

- Giving birth to live young after a period of protected development makes more demands on the mother but it provides a more secure and nutritious world for the emerging offspring than an amniotic egg laid immediately after fertilization and protected only by a hard shell.

- Gestation within the womb permits the development of a more complex brain and other structures that set mammals apart from all the other animals. That is why the brains of mammals can grow larger and more specialized than the brains of reptiles.

- After they are born, newborns are nourished with milk, a wonderful nourishing food unique to mammals. No other animals make food like this to meet the needs of their vulnerable young.

Chapter 8: Dinosaurs' Extinction – *Chalky Crash*

K-T Boundary Extinction

65 million years ago

It is called the "K-T Boundary Extinction" because it marks the end of the Cretaceous Period or era of chalk, known as Kreide in German, and the beginning of the Tertiary period. It wasn't the worst but it may be the most famous extinction because it killed the dinosaurs which had seemed impregnable for so long. Mysteriously, many mammals and birds were relatively unscathed.

Ten years ago, most scientists thought this extinction was caused by a meteor impact in Mexico. Now some think this particular meteor didn't land until 300,000 years after the last dinosaur had died.

Today many scientists think that it may have been the changing climate. Dinosaurs expelled huge amounts of methane, a gas even more potent than CO^2. Like the bacteria devastated two billion years before by the climate change created by their own pollution, it is possible that the dinosaurs themselves created the environmental change that eventually killed them.

The implications for the human race give pause for thought.

Cretaceous Period.....145 – 65 mya

The Dinosaurs: *Millions of Years Ago*

Writers have described Earth 65 million years ago immersed in a fiery inferno immolating everything that could not shelter under a rock or burrow in a hole. They may be rare, but when a meteor hits Earth, the impact can be devastating.

Asteroids gouging out craters as big as 300 feet hit Earth every 200-400 years. But serious meteors only arrive about every million years so we don't have first hand experience of what it's like to be here in such circumstances. What we do have is a view through the Hubble space telescope of a cosmic collision in 1994 when a meteor shower hit the planet Jupiter. Based on the battering Jupiter, a thousand times bigger than Earth, endured, scientists have speculated on the havoc following such a meteor strike here.

- A meteor blasts out craters far greater than its own size. It not only will kill everything it hits but will also start a fiery whirlwind, boiling the water in rivers and lakes, and possibly destroying all the forests on the continent. It could throw up a cloud cover severely reducing world sunlight and lowering temperatures for years. Rain could resemble the corrosive acid of Earth's first age.

- A meteor shower might bombard Earth for another week, creating further havoc by initiating giant ocean waves washing over miles of land, and causing volcanic explosions which, in turn, would ignite further fires and throw more toxic ash into the atmosphere.

A meteor impact leaves a layer of telltale ash darkened by iridium whose concentrations are much higher in extra-terrestrial rocks than in Earth's native rocks. Whether it is what killed the dinosaurs, we don't know. But we do know that something left a layer of iridium-soaked clay here on Earth 65 million years ago.

We don't know if it killed the dinosaurs or if they were already extinct. But there is no question that it did a lot of damage.

Chapter 8: Dinosaurs' Extinction – *Chalky Crash*

Chunk III

The Age of Mammals Begins

65 Million Years Ago

The Age of Mammals ushers in the third and most recent eon of life on Earth. It began with the extinction of the dinosaurs.

During the long reign of the dinosaurs, mammals had remained small, surviving in the niches too insignificant for the mighty beasts that strode the land. But mammals were able to survive in those small niches when the dinosaurs were obliterated in a strike and it is now the mammals that dominate almost every corner of the Earth.

Many of the animals familiar to us today – the horse, the pig, the dog – evolved shortly after the dinosaurs' extinction. It is the primates, however, who began to walk on two feet, and who gave birth to the first species of humans several million years ago. These humans were not yet *Homo sapiens,* but they emigrated out of Africa into many far reaches of the globe and by at least half a million - 500,000 - years ago had control of fire to cook food and to keep warm.

9. ✦ 65 million years ago: At the end of the Cretaceous Period and the extinction of the dinosaurs, the age of mammals began. New mammals evolved in many different forms, some which we recognize today, many more which have not survived.

10. ✌ 7 million years ago: When some members of a group of mammals called primates began to walk upright, using two feet instead of four, a new way of life began to emerge.

11. ☝ 1.8 million years ago: The first humans belonged to the genus but not the species of *Homo sapiens*. Many human species which are now extinct developed since then.

Chapter 9

Mammals Replace Dinosaurs

65 to 23 Million Years Ago

Age of Mammals Begins: *65 Million Years Ago*

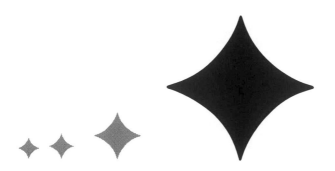

Today's Mammals

Between 65 and 23 million years ago, almost all our familiar and many now unfamiliar mammals evolved. The climate, initially warm and dry, gradually grew cooler, and repeated bouts of glaciation ultimately made life for the some of the largest mammals unsustainable. But the smaller mammals we know today survived the cold. New fruiting plants and grasses also flourished, providing food for the energy-hungry mammals.

Perhaps most momentously, the first primates evolved 60 million years ago.

Mammals Grow Up

65 million years ago

Mammals, after millions of years of subdued existence, moved into the dominant positions vacated by the dinosaurs in less than three million years. By 50 million years ago, some of the largest land mammals that have ever existed had evolved, but many species we know now were miniatures then. Their very smallness might be why many survived. The warm weather of the last 200 million years was coming to an end and the smaller animals were often better able to adapt to the cold. Many of the largest and fiercest mammals died out.

Paleocene Epoch   65 - 56 mya

Age of *Mammals* Begins: *65 Million Years Ago*

more about…
modern mammals

The ancestors of most mammals we know evolved at this time.

- The semi-aquatic ancestor of whales, dolphins and – surprisingly - the hippopotamus- evolved as herbivores. When mammals like whales first returned to the sea, they grew to no more than six feet.

- The horse family appeared as 24" miniatures.

- Bats evolved wings from hands used by earlier mammals to climb.

- Elephants grew to a maximum of two feet.

- Members of the cow family evolved later than horses, and have special stomachs to break down the hard cell walls of plants and evolved later than horses. They include the pig, camel, deer, llama, giraffe, sheep, wild boar, and of course, cows. They also were very small. Deer were the size of today's rabbits.

- The cat family included what have become today's domestic cat, the hyena and mongoose.

- In the dog family, ancestors of wolves, sea lions, bears, walrus, otters and racoons appeared.

Chapter 9: Mammals Replace Dinosaurs –
Today's Mammals

First Primates

60 million years ago

Proto primates or prosimians appeared quite soon after the dinosaurs died. They were originally tree dwellers and their distinguishing features were helpful for living a high life. They had flat nails instead of claws, opposable thumbs and toes for getting a good grip, binocular vision, and were less sensitive to smell than ground-hugging animals. They developed a brain with a posterior lobe which increased their information processing capacities, and their skeletons made them accomplished at the acrobatics handy for tree travel.

If some of these characteristics sound familiar, you are right. These are our very ancient ancestors.

Paleocene Epoch   65 - 56 mya

Age of *Mammals* Begins: *65 Million Years Ago*

more about...
primate lineage

Palaeontologists have not found "the mother of all modern primates," which remains one of the many missing links in the tree of life. But here is the long ancestral line leading to the primates as we know it.

Million years ago:

- *4,000 mya:* the first single-celled organism on Earth
- *1,700 mya:* eukaryotes – single complex cells including

pre-animals using respiration to generate energy

- 650 mya: flat worm - developed bi-lateral symmetry
- *545 mya:* pikaia – early chordate; a pre-vertebrate with an incipient backbone and front-end nerve swelling
- *500 mya*: fish –the first true vertebrates with an internal skeleton and backbone
- *370 mya:* tetrapods – amphibians, fish that migrated to land life and developed four limbs
- *300 mya:* synapsids – reptiles with strong jaws and a skull hole that made it possible to breathe and eat at the same time
- *260 mya:* therapsids – half reptile, half mammal, an evolutionary bridge between them
- *240 mya:* mammals – tiny, shrew-like warm-blooded animals evolved along side the first dinosaurs
- *125 mya:* eomaia – an early mouse, the first known placental mammal
- *60 mya:* the first primates or prosimians

Chapter 9: Mammals Replace Dinosaurs –
Today's Mammals

Modern Mountains

50 million years ago

About 50 million years ago, the clashing of continents began to push up Earth's younger mountains. The Pyrenees resulted from a collision of land masses we now call Iberia and France. Land we call Italy today created the Alps by moving north. The Rockies and the Andes formed during this time, and India's run-in with Asia is still pushing up the Himalayas and the high Tibetan plateau.

Eocene Epoch   55 – 34 mya

Age of *Mammals* Begins: *65 Million Years Ago*

By studying rock formations, geologists can tell where the continents have been for at least 700 million years.

• 700 million years ago the continent <u>Northern Gondwana</u> included India, Australia, and Antarctica; <u>Southern Gondwana</u> contained Africa, both American continents and most of Asia.

• 550 million years ago the continent <u>Laurasia</u> was made of North America, Scotland, and Greenland. Along with <u>Gondwana</u>, it was one of the world's two mega-continents for 250 million years. <u>Avalonia</u>, a continental fragment covered by the sea, included England and Wales.

• 300 million years ago, the supercontinent <u>Pangaea</u> formed when <u>Gondwana</u> and <u>Laurasia</u> collided.

• 200 - 150 million years ago, <u>Pangaea</u> slowly began to split into Africa, the Americas, and Asia. The west coast of Africa and east coast of South America still look as if they could be pushed back together like an ill-fitting jig-saw puzzle.

• 120 million years ago, the Atlantic Ocean began to come between Europe and North America. Our familiar continents and oceans were generally in place by 65 million years ago.

• 35 million years ago, Australia was pushed into a tropical zone, Antarctica was isolated over the south pole, and Hawaii, Iceland, and Japan had all been created as volcanic islands.

• Today the continents may be re-converging.

Chapter 9: Mammals Replace Dinosaurs –
Today's Mammals

Grasslands and Prairie Animals

35 million years ago

35 million years ago, the long-term cooling trend began to change plant life. Oaks, conifers, and maples replaced northern tropical vegetation and new grasses swept across prairies, savannahs and steppes. Grazing mammals like giraffes, antelopes, and ancestors of the gazelle, buffalo, bison and yak fed on the lush grasses along with smaller animals like beavers and rabbits.

10,000 years ago, these grasses became the foundation of the farming on which modern civilizations depend.

Oligocene Epoch   34 – 23 mya

Age of *Mammals* Begins: *65 Million Years Ago*

more about…
our cool world

We are worried about global warming and so it is quite astonishing to most of us to discover that Earth is actually in an interglacial period of an ice age that began 10 million years ago. Here is how it happened.

- *55 mya:* the climate began to cool.

- *30 mya:* the drift of the continents over both poles kept warm equatorial air from circulating around the world. Climates became cooler and drier; forests shrank, and deserts and prairies expanded.

- *10 mya:* permanent ice covering the south polar region marked the beginning of the present ice age.

- *6 mya:* the Mediterranean locked up 6% of the salt in the world's oceans. This raised the freezing temperature and polar ice expanded, making it even colder.

- *3.5 mya:* ice sheets covered both polar regions. They have remained ever since, even when the global climate entered warmer interglacial periods.

- *13,000 ya:* the glacial cold gave way to the present warmer interglacial. Today, maybe for the first time in Earth's history, polar ice caps have begun to melt during an interglacial of an ice age as a result of warming produced by greenhouse gases. Yet at this point, Earth is still cooler than usual. Usual, that is, in global terms encompassing Earth's climate for the last 4 billion years.

Chapter 9: Mammals Replace Dinosaurs –
Today's Mammals

Chapter 10

Walking on Two Feet

23 to 2 Million Years Ago

Age of *Mammals* Begins: *65 Million Years Ago*

Intimations of Humanity

Several pivotal things happened during this time. Each occurred during a period of global climatic upheaval and they all happened in Africa.

First, the superfamily called hominoids evolved about 23 million years ago. Hominoids weren't human – they weren't near being human – but they were the patriarchs of the family into which *Homo sapiens* ultimately was born.

16 million years later, hominines descended from the trees to walk on two feet.

Finally, stone tool-making emerged, made possible by a larger brain and two free hands which are better than feet for this kind of thing.

The Hominoid Superfamily

23 million years ago

23 million years ago, the superfamily of hominoids, the most intelligent man-like apes, appeared in Africa. They were immensely adaptable and by 18 million years ago they were living in Europe and Asia. Today the hominoid superfamily includes chimpanzees (our closest known living relatives), gorillas, orang-utans, gibbons and *Homo sapiens*, who, as far as we know, are the only surviving human species.

At first, scientists thought our evolution could be traced straight back to the great apes. That was before we began to find the evidence. Now scientists often don't know into which category a fossil may belong or by what circuitous route it may be related to us or its contemporaries. Instead of a straight line, the road back to our roots looks more like a bush with a thicket of offshoots. At least we know now that there's a lot we don't know.

Miocene Epoch   23 – 5 mya

Age of *Mammals* Begins: *65 Million Years Ago*

more about...
our family muddle

Theoretically our ancestors should all fit neatly into the primate classification. But about 20 years ago, scientists realized their definitions of hominoids and hominids didn't fit the facts. So they changed the definitions. But disagreement and confusion still reign. Bi-pedal primates, who are ancestors to the homo genus, used to be called hominids but now are called <u>hominins</u> and the broader term <u>hominid </u>is used for bi-peds along with the great apes who do not usually walk on only two feet. Should you want to know, here is the place where *Homo sapiens* presently stands in one of the current scientific scheme of things.

<u>Kingdom:</u> animals - organisms that use respiration

  <u>Phylum:</u> chordates – all vertebrates, animals with a backbone

    <u>Class:</u> mammals – groups whose females nurse their young

      <u>Order:</u> primates – the most highly developed mammals

        <u>Superfamily:</u> hominoids – great apes, orangu-tans, and gibbons

          <u>Family:</u> hominids –chimps, gorillas and bi-peds

          <u>Subfamily:</u> hominines –all bi-pedal primates

            <u>Tribe:</u> hominins – bi-pedal primates who are our ancestors

              <u>Genus:</u> *homo* who evolved as various species

                <u>Species:</u> many *homo* species have been identified including *Homo sapiens, ergaster, erectus, neanderthalis* and *floresiensis*. How many of these are really distinct from one another and which have simply been given different names is not yet clear.

Today *Homo sapiens* is the only surviving species known to us in the tribe of hominins. New evidence offers the possibility that other hominins lived as recently as the last century and may yet be living in parts of the world to which we have not penetrated.

Chapter 10: Walking on Two Feet –
Intimations of Humanity

119

Primates Walk on Two Feet

7 million years ago

The first primates who walked on two feet lived in Africa. Why this two-foot fashion caught on is a source of some scientific debate. It leaves two hands free and hands are better than feet for writing, cooking, or playing golf, but the evolutionary advantage 7 million years ago is not so clear. You can't run as fast on two legs as on four, and it's harder to stand up straight.

Nonetheless, once it got started, walking upright was a popular innovation and as many as 20 or 30 bi-pedal species may have appeared and disappeared throughout Africa between 7 and 3 million years ago. Among them emerged the tribe of hominins, our ancestors, though we may never know who most of them were.

Miocene Epoch   23 – 5 mya

Age of Mammals Begins: *65 Million Years Ago*

more about...
reading old bones

Fossils don't come with little identifying tags, so scientists do a lot of detective work before they know what they have found. One of the most well-known bi-pedal fossils is Lucy, more formally known as *Australopithecus afarensis*, who lived in Ethiopia 3.6 million years ago. How do scientists know how many feet Lucy walked with?

There are several distinguishing characteristics in the bone structure of bi-peds. Their arms are usually shorter relative to the length of their legs, and their thumbs but not their toes are opposable. The most critical feature is the entry point of the spine into the skull, which is lower in bi-peds than in four-legged walkers.

Yet it is not as easy as it may sound and it is a puzzle even to know conclusively how many different human species have existed. A species is a group of organisms which can interbreed. Members of the same genus may have common ancestors but cannot interbreed with a different species even if they belong to the same genus.

Perhaps all the humans which have ever existed belong to a single species. Most scientists, though, think there were many different human species which have died out. Just how extensively *Homo sapiens* interbred with other human species is unclear. It is possible that genes, particularly from male lines, have been lost when conquering "alpha males" demanded exclusive sexual rights to females from the conquered tribes.

The history of palaeontology includes many misinterpretations and even a few old treasures which have been tossed away as trash. There are also cases of successful fraud. The most famous is the skull of Piltdown man, on display in London for years as the missing link between ape and man, until it was revealed to be a man-made construction of bones from more than one species.

Research and debate will no doubt continue for a long time yet.

Chapter 10: Walking on Two Feet –
Intimations of Humanity

Stone Tools

2.5 million years ago

We used to think that tools were unique to humans. We know today that many animals including birds make use of tools. Stone tools have left a record of their use, however, that makes them unique.

Archaeologists have found what we would probably call a tool factory in Olduvai Gorge, Tanzania, Africa. It's between 2.9 and 2.5 million years old, and is the oldest evidence ever discovered of the systematic manufacture of stone tools. They must have worked pretty well because their design remained basically unchanged for a million years.

Currently, the best guess is that the tools were first crafted by *Homo habilis*, or "Handy man," whose skill enabled him to add meat to his diet. This nutritional change provided valuable fuel for a brain 50% bigger than a modern chimp's, and almost certainly contributed to the eventual development of humans' superior intellect. Although *Homo habilis* showed greater intelligence than other primates, and was first given the status of early man, most scientists now consider this mainly an honorary title, and think he did not yet belong to the *Homo* genus.

Pleistocene Epoch   2.6 mya - 13,000 ya

Age of Mammals Begins: *65 Million Years Ago*

more about…
the story in the stones

Tools made of softer materials may have been made millions of years before stone tools. But it is the stone that survived over time and the tools tell a fascinating story about the human brain.

● These earliest known tools may appear primitive, but no non-human primate living today, even with training, can produce implements as sophisticated as these. Stones had to be chosen carefully and hammered precisely. They distinguished between right and left hand use, so they seem to reflect the beginning of a bi-lateral brain.

● These early tools were made and then used later in different places. *Homo habilis* planned ahead, and carried his tools with him. How simple, how obvious this seems. Yet no other primate would do such a thing today. Even chimpanzees don't carry the stones around they may spontaneously think of using. The tools found in Olduvai Gorge required forethought, made possible by a neocortex, what eventually becomes the largest part of the human brain.

● The neocortex also creates mental maps which would have helped *Homo habilis* broaden his hunting territory, and improve both his diet and his competitive edge.

● Finally, cooperative social behaviour is closely related to the size of the neocortex. The makers of these tools were capable of working together in a social world of increasing complexity.

Chapter 10: Walking on Two Feet –
Intimations of Humanity

Earth Trembles

2 million years ago

Some strange things happened two million years ago. Supernova explosions in our solar system's neighborhood, the first five and another two million years ago, may have shaken Earth to its core. It is probable that they also created the Local Bubble, an unusual, almost empty, space that surrounds us and makes it possible for us to see the stars. Nobody knows for sure if the explosions caused the north and south magnetic reversals that also occurred at these times. Or if they caused the drastic change in the weather or explain a series of sudden extinctions that left room for human evolution to continue.

Whatever their causes, from a human point of view, it was the beginning of a new period in Earth history.

Pleistocene Epoch   2.6 mya - 13,000 ya

Age of *Mammals* Begins: *65 Million Years Ago*

more about…
climatic gyrations

For some thus far unknown reasons, climatic gyrations during the last two million years have been uncharacteristically extreme. Two million years ago, a glacial lasting 90,000 years was followed by a warmer interglacial lasting about 10,000 years. Since then, this pattern has been repeated every 100,000 years or so. Ice sheets, rain forests and savannahs have been shrinking and expanding, and rising and falling sea levels have exposed land and covered it again, creating land bridges that have connected and then disconnected continents.

Given this pattern, we would expect our current interglacial warming to be coming to an end. Instead, it looks as if our climate will get warmer - perhaps even dangerously warmer - as a result of greenhouse gases emitted by human activity.

Pivotal evolutionary developments have most often occurred during periods of abrupt environmental changes, and these climatic alternations may have been an important impetus for the unprecedented evolution of the human species. We don't know if we ourselves might be engineering something similar for the future.

Chapter 10: Walking on Two Feet –
Intimations of Humanity

Chapter 11

The First Humans:

1.9 - million years ago

Age of *Mammals* Begins: *65 Million Years Ago*

Homo but not Sapiens

1.9 million years ago, people belonging to the *Homo* genus evolved. They had bodies with human proportions and complex bilateral brains two-thirds the size of modern humans. Greater equality between males and females reduced sexual competition and led to a more influential role for fathers during the long years of human childhood.

The First Humans

1.9 million years ago

Although it may not be clear whether *Homo habilis* really belonged to the *Homo genus*, *Homo ergaster* or "workman" and *Homo erectus*, his later relative, clearly did. Their brains were two-thirds the size of ours and they had tall lean bodies. Their hand axes, spears, and containers, called "Acheulean," are more sophisticated than earlier stone tools, they constructed rudimentary dwellings, and their social relations were more cooperative. We don't know if they could talk – possibly they could - but the evidence suggests conclusively they were human, belonging to the same *Homo* genus to which we belong.

Ergaster and *erectus* were able to adapt to diverse habitats and were the first hominins to leave Africa, moving through Eurasia and China. Scientists called his skeleton *Java Man* when it was found in Sumatra, Peking Man in China, and he may be *Homo georgicus* in Georgia (in the ex-U.S.S.R.). He was so spectacularly successful that *erectus* may have been the only surviving hominin for a million years and some scientists think *Homo erectus* is the founding father of all later humans, including us. Other scientists think the lines of *ergaster* and *erectus* diverged and that *sapiens* and possibly other human species descended from *ergaster* but not *erectus*. We just don't know yet.

Pleistocene Epoch   2.6 mya - 13,000 ya

Age of *Mammals* Begins: *65 Million Years Ago*

more about…
the human family

Homo *ergaster* and *erectus* belong to the genus of *Homo*. What made them different from their predecessors was a bigger and more complex brain. Essential to human braininess was the emergence of a long-term family.

The size of the mother's birth canal limits the size of a newborn. So among humans, most of the brain's development occurs after birth and offspring need care for much longer than the offspring of other animals. Because the birth process is more arduous for the human mother and parenting lasts so much longer, both mother and child benefit if the number of offspring is limited, giving both the nurturance they need. At the same time, the value of a long term family is greatly enhanced.

The uniquely human family emerged in quite a surprising way. *Homo* males and females became more alike and more equal. Males were no longer significantly bigger than females and the oestrous cycle, which made it physically apparent when a female was potentially fertile, disappeared. Both males and females became sexually active even when conception was impossible.

This partially separated sexuality from reproduction and broadened the foundation of relationships. Outright physical competition among males for female sexual favors gave way to pair-bonding. Males acquired a long-term sexual partner, provided support for mothers and their offspring, and took on a more significant parenting role.

Chapter 11: The First Humans - *Homo but not Sapiens*

Controlling Fire

1 million years ago

We're not sure when man first learned to tame fire since proof of its use, for obvious reasons, tends to be sparse. Their mere survival in areas of intense cold makes one think people might have been able to warm themselves with fire as early as 1.5 million years ago. 600,000 years ago, a series of glacial periods occurred along with what seems to have been a rapid spurt when human brains reached 90% of their present size. There are signs that people learned to use and even ignite fire then. There is incontrovertible evidence of hearth fires 300,000 years ago.

Some anthropologists think control of fire began much earlier, close to two million years ago and around the time the first humans began to craft stone tools. It was also the time when the human brain got bigger, and the guts, mouth, teeth and jaw of humans became smaller. This change suggests that we no longer needed the large teeth and strong jaws required to consume hunks of uncooked meat. Eating cooked instead of uncooked food would also have released significant amounts of energy formerly needed for digestion. That would make increased energy available for supporting a larger brain and its greater capacity for thinking.

Pleistocene Epoch   2.6 mya - 13,000 ya

Age of Mammals *Begins*: *65 Million Years Ago*

more about...
the technology of fire

Making things is not uniquely human. Beavers raise dams, spiders weave webs, bees construct honeycombs, birds build nests. These are complex feats. But no other animal has ever achieved control of fire. In the history of human technology, with its seemingly limitless, unstoppable, adaptability, it is hard to imagine any other technological triumph that has changed living conditions so fundamentally as harnessing fire.

● Fire gave humans control over heat and light, freeing us from the tyranny of the sun's inflexible rhythms.

● Fire made it possible to keep animals out of the caves humans wanted to inhabit themselves.

● Control of fire extended human habitats into colder regions.

● Ten thousand years ago, fire was used to clear fields for planting. It improved the manufacturing of tools and advanced the use of metal implements.

● Quite possibly the most significant contribution of fire occurred when man began to use it to cook food. Cooking makes food safer to eat by killing parasites and bacteria. Cooked food doesn't require as much energy to digest, and so it frees up time that must be used by other organisms to process what they eat.

If cooking food began as long ago as two million years ago, it could explain why the brains of *Homo erectus* were 50% bigger than their ancestors. The fire used to cook food, therefore, may be a key component in the difference between being an advanced ape and an early human.

Chapter 11: The First Humans - *Homo but not Sapiens*

First Humans in Europe

.8 million (800,000) years ago

A startling find of fossils deep in an Iberian cave shows that people arrived in Europe at least 800,000 years ago. They are called *Homo antecessor*, meaning "pioneers," and by 700,000 years ago were living as far north as Suffolk, England, where archaeologists who discovered his fossils have called him *"Lowestoft Man."* We don't know what happened to *antecessor*. Perhaps they died out during an extreme glacial period. By 500,000 years ago, their places in Europe had been taken by *Homo heidelbergensis*, the ancestors of *Neanderthal man*.

Pleistocene Epoch   2.6 mya - 13,000 ya

Age of *Mammals* Begins: *65 Million Years Ago*

more about…
threatened primates

The great apes, that is gorillas and chimpanzees, share more than 96% of their DNA with humans. Some scientists say chimps are so similar to us they should be classed as a human species.

But unlike us, great apes need an undisturbed forest to survive. Today their habitat is being destroyed and the great apes are dying at an alarming rate. Logging is destroying their forest homes, hunting them for sport or food is killing them faster than they can breed, and disease in some areas is devastating them.

If nothing is done, scientists believe that by 2030, there will be almost no habitat left that is suitable for these wonderful intelligent beings. They will no longer live anywhere save in the squashed imprisonment of occasional zoos. It would be a terrible cruelty and an irreplaceable loss.

Chapter 11: The First Humans - *Homo but not Sapiens*

Chunk IV

First Age of *Homo Sapiens*

250,000 to 13,000 Years Ago

Homo sapiens, meaning "wise man," is the species we have named ourselves.

The strongest evidence is that *Homo sapiens* evolved in Africa about 250,000 years ago and emigrated out of Africa at least 80,000 years ago. It is possible, though, that *Homo sapiens* evolved from the earlier *Homo erectus* who may have emigrated out of Africa almost two million years ago.

12. ☝ 250 thousand years ago: *Homo sapiens* first evolved in Africa.

13. 🏹 80 thousand years ago: This may have been the first or merely the most recent successful mass emigration of Homo *sapiens* out of Africa into China, Australia, and Papua New Guinea.

Chapter 12

Homo Sapiens

250,000 to 13,000 Years Ago

First Age of Homo Sapiens: *250,000 - 13,000 Years Ago*

Our Beginnings

Homo sapiens probably evolved in Africa about 250,000 years ago. So we have been here for about 5% of the all the time Earth has been in existence. In this relatively short time, the changes which we have wrought, for better and for worse, have been staggering.

Homo Sapiens

250,000 years ago

Homo sapiens, the species to which we belong, were probably evolving by 300,000 years ago. We know from fossils that our distinct physical anatomy was clearly established by 200,000 years ago though we may never know exactly when a mind we would recognize as clearly human evolved.

Currently, the most popular theory is that "Eve," not actually a single woman but a core of between 2,000 and 10,000 Africans, lived some 200,000 years ago. Our "Adam" also was not literally a single male progenitor.

Pleistocene Epoch   2.6 mya - 13,000 ya

First Age of Homo Sapiens: *250,000 - 13,000 Years Ago*

more about…
when did we start?

Despite our uniqueness, it is difficult to say precisely when *Homo sapiens* evolved. Many scientists think our evolution took place gradually over hundreds of thousands of years and that there may be no single event that unambiguously marks our arrival. Other scientists think a significant genetic change, possibly quite recently, will eventually prove to mark the definitive advent of modern humans.

We do know that for millions of years, there were people like *Homo erectus* and *Homo ergaster* who walked on two feet, had large forebrains and spent many years rearing their offspring. They hunted in groups, used tools, and stalked prey. They wore clothing, looked after their sick, and buried their dead. It is very likely that they had linguistic skills. They emigrated out of Africa some two million years ago.

Did they belong to the same human species we do? How one answers this question may depend as much on philosophical values as on what may always be inconclusive evidence.

It is rather like our understanding of the Big Bang. There is so much we do know that it is amazing to discover how much more there is that we don't.

Chapter 12: Homo Sapiens - *Our Beginnings*

Abstract Thought

200,000 years ago

We who belong to *Homo sapiens* have a lot in common with other animals. Yet we are extraordinary. We read poetry and celebrate birthdays. We care who wins a football game and we write wills to distribute our property when we die. We've gone to the moon and are trying to get to Mars. Why? What makes it possible for us do these things?

Scientists have been trying to figure out when our ancestors started thinking the way we think today. The advanced level of our abstract thought and symbolic language seem to be the twin abilities that make us unique. But these abilities are hard to pin down. Abstract thought may be much older than the artifacts we have so far unearthed or that have even been preserved at all. Or we may misinterpret what we do find. Abstract thought of some kind may have begun several million years ago and evolved gradually. Or abstract thought as we know it, may have erupted suddenly with a critical genetic change as late as 50,000 years ago. Everybody is still making pretty big guesstimates.

Pleistocene Epoch   2.6mya - 13,000 ya

First Age of Homo Sapiens: *250,000 - 13,000 Years Ago*

more about...
human thinking

Thinking and talking didn't leave direct artifacts like tool-making did. So in the search for our roots, scientists examine fossils to see if people were anatomically similar to modern humans. They also look for artifacts that could be crafted only by people able to manipulate mental images and think about things beyond their immediate experience. Abstract thought is reflected in art, in various social interactions, in some kinds of tool-making, and problem-solving

Abstract thought may have developed rapidly with a single genetic change. But there is intrigueing evidence that it emerged gradually. Most of the earliest artifacts have been found in Africa.

<u>years ago:</u>   <u>people had made</u>

- 2,500,000 - what seem to be art-like objects.

- 400,000 - what may be a human figure carved of quartzite.

- 250,000 - standardized, precisely-made tools.

- 200,000 - rock art carved into cave walls in central India

- 140,000 - regular long distance trades requiring complex social networks and long-term goals.

- 130,000 - hand-made pendants suggesting concepts of identity, beauty, and social status.

- 77,000 - carvings and jewelery of carefully drilled shells possibly used as an early form of money.

The evidence is, not surprisingly, sparce and what we have was left by people living in cultures lost forever. We will never know with certainty if modern thought is intrinsically the same as the thinking of people who hunted wild animals with hand-made tools. How and when true abstract thinking really started may always be one of the more controversial issues in science.

Chapter 12: Homo Sapiens - *Our Beginnings*

Neanderthal Man

125,000 years ago

When *Homo sapiens* were evolving in Africa 200,000 years ago, *Homo neanderthalenses* were evolving in Europe. Initially, there seems to have been little difference between the two groups and in 1856 when their fossils were found in Neander Valley in Germany, many thought we belonged to the same species. Although Neanderthals are sometimes portrayed as backward, club-wielding thugs, they had brains as large and complex as ours, and would have been capable of the same levels of abstract thought. They could talk and sing, and are responsible for some of the ancient cave art found in Spain. They cared for their sick, and buried their dead. Their tough stocky bodies gave them immense physical stamina and endurance against extreme cold.

For more than 165,000 years the Neanderthals occupied Europe, spreading from Spain to the borders of China and as far north as the Ukraine. They were a human species and co-existed with *Homo sapiens* in Europe for up to 20,000 years. Yet, 25-30,000 years ago, all trace of the Neanderthals disappeared.

Pleistocene Epoch   2.6 mya - 13,000 ya

First Age of Homo Sapiens: *250,000 - 13,000 Years Ago*

more about...
what happened to our fellow humans?

The extinction of the Neanderthals makes the question of our own humanity particularly anguishing. Why did they disappear?

Perhaps the Neanderthals -

- ...interbred with *Homo sapiens*, and we are now members of one indistinguishable family. Recent analysis shows that 2 to 4% about 2% to 4% of the DNA of individual Europeans is Neanderthal,

- ...did not realize how murderously aggressive *Homo sapiens* could be. Perhaps our species hunted them to extinction as we have hunted other species and are now threatening our remaining primate relatives. Yet, there is no evidence, no skeletal remains suggesting that *Homo sapiens* slaughtered the Neanderthals.

- ...died out because their hunting skills failed them. They disappeared after a period of gyrating weather extremes that destroyed many of the woodlands where they were expert hunters. Their bodies gave them strength but they were not agile runners, and they may have starved when animals could flee across open plains.

- ...lacked sufficient genetic diversity as a species to survive the kind of pandemic which has through the centuries devastated our own species.

At the time *Homo neanderthalis* became extinct, *Homo sapiens* were learning to trade and diversify. Those who were not hunters made clothes and utensils, and developed useful skills that could be exchanged for food. It could have been individual differences and free trade that saved *Homo sapiens*.

Chapter 12: Homo Sapiens - *Our Beginnings*

Chapter 13

Going Global

80,000 to 13,000 Years Ago

First Age of Homo Sapiens: *250,000 - 13,000 Years Ago*

We Colonize the World

Homo sapiens migrated out of Africa about 80,000 years ago. But it is not clear whether this was the first migration and if all non-Africans today are descendants of this group. The evidence does not rule out the possibility that there were repeated migrations and inter-breeding among various human species in Asia, Europe, and Africa for as long as 600,000 years.

However it occurred, and despite a devastating supervolcano that seems to have reduced human population to no more than several thousand, our species was successfully established as far east as China and as far south as Australia and Papua New Guinea 70,000 years ago. By the time the most recent ice age forced a withdrawal from central and northern Europe 25,000 years ago, human culture had spread throughout at least half the globe.

Out of Africa

80,000 years ago

Many scientists think *Homo sapiens* lived in Africa for over a hundred thousand years and that a small band left Africa 80,000 years ago. They believe this was the only successful exodus by *Homo sapiens* out of Africa, and all non-Africans - Asians, Europeans, Indians, Orientals, Australians and Native Americans - are descendants of this single pioneering group.

Research, however, suggests that this simple story may not fully account for what really happened. In fact, at this point, DNA analyses and fossil finds far from Africa offer a tantalizing puzzle which we have not yet completely pieced together.

Pleistocene Epoch   2.6 mya - 13,000 ya

First Age of Homo Sapiens: *250,000 - 13,000 Years Ago*

more about…
our treks out and about

Our species today has migrated to almost every nook and cranny on Earth. How and when did this happen? There are three principle possibilities.

The Out of Africa or Replacement Theory

According to this theory, modern humans evolved in Africa between 100,000 and 200,000 years ago. Probably about 80,000 years ago, they migrated out of Africa eventually reaching Asia, Australia, Papua New Guinea and Europe. Any other human species, like the Neanderthals, were replaced by *Homo sapiens* who today are the only surviving humans anywhere in the world.

Current fossil records support this view. The earliest *Homo sapiens* skeletons found so far come exclusively from Africa beginning about 200,000 years ago.

This sounds convincing until one looks at recent research. Both DNA and skeletal analyses indicates that there has been a genetic contribution from other archaic humans besides *Homo sapiens*. The finding that the gene betaglobin which appeared some 200,000 years ago is more widely distributed in Asia than in Africa is one of the most problematic discoveries. It doesn't support the view that *Homo sapiens* emigrating out of Africa 80,000 years ago replaced humans already living there.

Chapter 13: Going Global - *We Colonize the World*

Multi-Regional Evolution Theory

A theory that overcomes some of the difficulties of the Out of Africa Replacement theory suggests that modern man evolved at more or less the same time in more than one place. Archaic humans, such as *Homo erectus*, for instance, evolved into *Homo sapiens* in Asia and locations in the Old World. At the same time, there was sufficient emigration and gene flow among the groups to prevent the evolution of distinctly separate regional species incapable of inter-breeding. In other words, although *Homo sapiens* developed separately in different parts of the world, over all there has been a single and continuous human species evolving worldwide since *Homo erectus*.

This theory explains some of the anomalies among *Homo sapiens* fossils found around the world that the Out of Africa theory has difficulty explaining.

Above all, it explains some physical similarities shared between archaic humans living in the region as long ago as 200,000 years ago and modern *Homo sapiens* who emigrated there. Modern Europeans, for instance, share relatively heavy brow ridges and a high angle of the nose with the Neanderthals East Asians today share a particularly shaped incisor possessed by *Homo erectus* but rarely found among Africans or Europeans. Australian aborigines seem to share dental traits with pre-modern people.

But it is possible that all these similarities are the result of interactions among members of *Homo sapiens*, all of whom originated in Africa. The findings do not conclusively rule out this possibility.

First Age of Homo Sapiens: *250,000 - 13,000 Years Ago*

Assimilation Theory

The Out of Africa and Multi-Regional Evolution theories both run into difficulties explaining all the fossil and DNA evidence that has recently emerged. By incorporating aspects of each of the earlier theories, Assimilation Theory may be able to resolve many of the apparent inconsistencies.

Assimilation theory hypothesizes that *Homo sapiens* did first evolve in Africa. As they migrated, however, they did not simply replace human populations like the Neanderthals already living there, but often interbred with them before the original populations were replaced.

This theory would explain why the oldest fossil evidence of *Homo sapiens* has been found only in Africa, but also why later fossils of *Homo sapiens* and why many of our own present-day characteristics show traces of archaic humans who are now extinct. This view is supported by DNA analysis which suggests that people of Africa, Europe, and Asia may have been interbreeding for as long as 600,000 years.

It also suggests that people have been emigrating out of, and possibly immigrating again into, Africa for hundreds of thousands of years.

Chapter 13: Going Global - *We Colonize the World*

The Toba Supervolcano

74,000 years ago

74,000 years ago, a cataclysm of almost unimaginable dimensions brought our species to the brink of extinction. A supervolcano in Toba, Sumatra, Indonesia erupted. If you look at a map, you can see that, given the size of a supervolcano's reach, it may have exploded in what was practically the epicenter of the human family.

It sent three billion tons of sulphur dioxides into the atmosphere and for at least five years, the world was plunged into a volcanic winter. Sunlight was blocked, and plants had no place to grow because deep ash blanketed the ground for thousands of miles. It was so terrible that the entire human population was reduced from over a hundred thousand to several thousand. It is possible that for the next 20,000 years, the number of people in the world never grew beyond this limited number.

Pleistocene Epoch   2.6 mya - 13,000 ya

First Age of Homo Sapiens: *250,000 - 13,000 Years Ago*

Shifts of tectonic plates cause earthquakes, tsunamis, and volcanoes. Supervolcanoes are like volcanoes only thousands of times bigger. On land or underwater, they can cause almost unimaginable devastation. Ordinary tame volcanoes throw up a cubic kilometre of fine pieces of magma ash. Breathing it is like inhaling fiery glass. Supervolcanoes throw up thousands of cubic kilometres of this ash dropping it hundreds of feet deep over a space the size of a continent. Dense clouds of sulphur dioxide surround the globe for years.

Will it happen again?
Yes. Tectonic plates will continue to collide. When Mt. Tambora erupted in 1815, global temperatures dropped 6^0 Fahrenheit, summer disappeared and crops failed. 117,000 people died. In 1883, the Krakatoa volcano killed 80,000 and obliterated an island. In 2005, tsunamis caused by a giant underwater upheaval in Indonesian waters killed hundreds of thousands of people, some as far as 2,000 miles away.

Where will it happen?
Supervolcanic hotspots exist around the world. One is under Yellowstone National Park. Oregon is in danger of tsunamis, and earthquakes are an ever present danger to hundreds of thousands living in cities like San Francisco and Tokyo, and in vast areas of the Middle East. Tsunamis are possible in Britain, and an eruption on the floor of the Atlantic Ocean could send water racing inland for miles along America's densely populated East Coast.

Will it happen again soon?
A supervolcano explodes somewhere about every 50,000 years. One niggling worry: for the last 16 million years, Yellowstone has exploded on average every 600,000 years. The last Yellowstone explosion was 630,000 years ago.

Will we have a warning that it's coming?
Possibly not.

Chapter 13: Going Global - *We Colonize the World*

China, Australia, and Papua New Guinea

70,000 years ago

Despite the Toba supervolcano, less than 6,000 years later, people had migrated as far east as China and as far south as Australia and Papua New Guinea. Perhaps they joined archaic human populations that had arrived before them. Genetic markers in the populations today suggest so.

In any case, a severe glacial period 70,000 years ago created so much ice that sea levels were drastically lowered, making a sea crossing from Timor to Australia just feasible. It was still an impressive accomplishment. The crossing would have been at least 40 miles long and required seaworthy watercraft not found anywhere else for possibly as long as 30,000 years.

Pleistocene Epoch   2.6 mya - 13,000 ya

First Age of Homo Sapiens: *250,000 - 13,000 Years Ago*

more about…
bones, stones, and genes

How do scientists chart the movements of man so long ago? What do they look for? What is the basis of their theories? There are three sources of evidence, each of which alone is incomplete but contributes to the whole picture.

Bones: skeletons are analyzed to find their age, the size of the brain, whether speech was physically possible, sometimes the person's diet and even their diseases. The ancient skeletons are compared with those of modern humans, whose skulls are rounded with less protruding brow ridges and less pointed chins. Skulls typical of modern humans begin to appear in fossils about 160,000 years old, with continued transitional forms appearing for another 50,000 years.

Stones: many artifacts are studied but stone tools are the most prolific because they deteriorate at a much slower rate than tools made of substances like wood. The stone tools show different methods and levels of sophistication, and as people carried their tool-making skills with them, it is possible to follow their trail for hundreds of thousands of years and for as many miles.

Genes: genetic analysis is a complicated business but DNA often makes it possible to trace the lineage of people today back several hundred thousand years.

Stratigraphy, Radiocarbon Dating, and Dendrochronology: these are all methods used to help estimate the date a site was used. Stratigraphy studies the layers of the Earth's crust where the site is located. Carbon 14 deteriorates in organic materials at a predictable rate. Radiocarbon dating can reliably measure the age of organic materials up to 40,000 years old based on the amount of carbon 14 they still contain. Dendrochronology uses tree-rings to estimate ages.

Despite the sophistication of the evidence, not everyone always agrees on what it means, and new discoveries inevitably raise the possibility of new alternatives.

Chapter 13: Going Global - *We Colonize the World*

Cro-Magnon Man in Europe

45,000 years ago

As the world warmed up, settlers of our species called Cro-Magnons arrived in Europe from the Middle East. They lived as semi-nomads with extensive trading networks, probably much the way we would live under similar circumstances. They could weave and sew clothes, build huts, storage pits and quarries. They buried their dead with highly developed rituals, so they seem to have had a theory of an afterlife.

The Cro-Magnons left behind a stunning array of artistic artifacts, including awesome cave paintings of life-sized animals in southern France and Spain. They made statues, carvings, engravings and musical instruments, and decorated themselves with jewelery and body paint. They had dancing and singing, and an immediately recognizable human culture.

Pleistocene Epoch   2.6 mya - 13,000 ya

First Age of Homo Sapiens: *250,000 - 13,000 Years Ago*

The artifacts of the Cro-Magnons were so advanced relative to anything they had seen before that some scientists at first believed they had found the birth place of modern humans. We know today that Africa, not Europe, is where *Homo sapiens* almost certainly first evolved. Yet scientists are still not sure why the sophisticated artifacts from this period seem to have appeared so suddenly. Perhaps it is because so much else from earlier time has been lost (or at least not yet found), perhaps some genetic change accounts for this sudden and dramatic output.

Or perhaps the explanation is our collective knowledge. It is possible that the human brain has not changed substantially for as long as 160,000 years. We may know much more than many of our ancestors not because our brains are better but because we can begin with the foundation of learning our ancestors have left us. To appreciate how much each of us benefits from what has been learned by others before us, think of how many things we could not do if we grew up totally alone. How many of us could make clothes like the Cro-Magnons? build a hut, hunt a deer or catch fish? or use the telephone, a television or the internet? Even simple things like talking or tying a shoe lace would be impossible.

Most of us don't grow up in isolation and legacies of language, art, science, law and social customs are passed on to us in the family, at school, on the street, through books, television, stories, rituals, songs, painting, even dance.

Without it, each generation would have to begin quite literally in the stone age. But each new generation doesn't have to reinvent everything. It is what Isaac Newton meant when he said, "we all stand on the shoulders of giants."

Chapter 13: Going Global - *We Colonize the World*

The First Americans

20,000 years ago

Who they were and how and when people first came to the American continents is still a puzzle with missing pieces. There is no question that people were living in America 11,500 years ago. But they may have arrived up to 20,000 or even an astonishing 40,000 years ago.

DNA and archaeological evidence suggest that the first Americans had ancestors from Japan, Australasia, India, and Europe. As many as four or five genetically distinctive peoples constitute what we call the first Americans or American Indians.

If the first Americans arrived in four or five different waves, they may also have arrived by different routes. There is strong evidence that at least some walked across a "land bridge" that appeared several times during recurring ice ages between 75,000 and 14,000 years ago. Hunters and their families may have followed animals that provided food, clothing, shelter, and even dung for fires. Some of these people may have continued to move south by walking along an inland ice-free corridor along the western coast.

It is also possible that people arrived by boat. People living along the shores of northeast Asia could have "island-jumped" to the Americans, especially during the last Ice Age when low water levels exposed many islands that are submerged today. They also may have arrived by boat from Europe. There are similar tool types used in France about 18,000 years ago, and tools found in America.

Pleistocene Epoch   2.6 mya - 13,000 ya

First Age of Homo Sapiens: *250,000 - 13,000 Years Ago*

more about...
unsettling our certainties

Science is constantly upsetting our fixed certainties, forcing us to re-think what we thought we already knew.

Until recently, almost everyone agreed that the Americas were first settled by Mongoloid tribes crossing from Siberia to Alaska about 13,000 years ago. Many scientists are still convinced of this. But other scientists now think that Europeans were settling in North America thousands of years earlier. This so contradicted what people believed that some scientists were accused of making dating errors, or even of deliberately fabricating the evidence. But corroboration has been accumulating:

• Flint blades made by the Solutrean people in the south of France 20,000 years ago are almost identical to blades found in the United States which carbon dating indicates were used 16,000 years ago.

• DNA studies of some American Indian tribes show traces of a genetic legacy coming not from Asia but from France.

• Many artifacts still used by traditional Eskimo culture are similar to Solutrean artifacts. Modern Eskimos have demonstrated how people using their traditional watercraft and seafaring techniques could have crossed the Atlantic Ocean from France to America some time during the last ice age.

• There is also tantalizing evidence suggesting that people may have reached Mexico 25,000 years earlier than scientists first thought. 180 human footprints left in volcanic ash at the bottom of an abandoned quarry in Mexico absolutely baffled scientists when radio-carbon dating showed they'd been left 40,000 years ago. How can these footprints at the edge of a dried out lake possibly have gotten there so long ago? Some scientists think they are impossible to explain and doubt the evidence. For others, the footprints suggest the astonishing possibility that settlers arrived on the West Coast from Japan and Pacific Ocean communities by employing advanced watercraft and extraordinary sea-faring skills.

Chapter 13: Going Global - *We Colonize the World*

The Little People

18,000 years ago

Archaeologists have recently discovered the skeletons of small hominins, whose mature height was no more than 3 feet, who were living on the Indonesian island of Flores until at least 18,000 years ago. They built fires, used stone tools, hunted small elephants and giant rats, fought off Komodo dragons, and could talk. Some scientists think the diminutive Flores people descended from *Java Man*, whose ancestor, *Homo erectus*, left Africa 1.6 million years ago. 900,000 years ago they were marooned on Flores, and in genetic isolation, evolved into a new species, dwarf versions of their predecessors. If we are descendants of *Homo erectus* in Africa, *Homo floresiensis* is our cousin.

Alternatively, it is possible that the Hobbits, as *floresiensis* is affectionately called, may have themselves emigrated from Africa, quite possibly the first hominin to do so.

A volcanic explosion 12,000 years ago may have caused the extinction of people on Flores but Indonesian tribesmen reported interacting with similar peoples further east up to a century ago.

Other reported sightings in remote regions are tantalizing. In 1951 an expedition to Mt. Everest took photographs of giant footprints in the snow which still remain unexplained.

Could folklore about elves, leprechauns, the Yeti, and fairies be based on real people still living in outposts we don't know about? Some scientists are taking the possibility seriously.

Pleistocene Epoch   2.6 mya - 13,000 ya

First Age of Homo Sapiens: *250,000 - 13,000 Years Ago*

more about…
a trail of extinctions

What will happen if we find other hominins still living somewhere today? Would coming in contact with our species be as deadly for them as it has been for so many others?

● Through aggressiveness, intelligence, ingenuity and ignorance, humans have changed almost every environment into which we have wandered. The relative importance of climatic change and human activities is not always clear. But we know many animals, especially mammals, became extinct shortly after our arrival, possible victims of over-hunting and diseases that arrived with us.

● Within several hundred years after humans arrived on the continents where they lived, many marsupials, land-dwelling crocodiles, giant kangaroos and armadillos, the mammoth, sloth, woolly rhinoceros, giant Irish elk and hundreds of other species, large and small, were gone.

● Today, many ocean-living species are under threat from over-fishing and pollution, and deforestation is threatening some of the largest and most intelligent animals still surviving in the wild.

Millions of species died and Earth underwent drastic environmental changes before humans ever appeared. So it is unrealistic to blame ourselves for every bad or good thing that has happened in the last fifty thousand years. Nonetheless, our survival depends on life forms as small as bacteria and as big as our grazing animals. We must find a way of preserving our biodiversity or we ourselves will perish.

Chapter 13: Going Global - *We Colonize the World*

Chunk V

Farming, Cities and Civilizations

13,000 - 600 Years Ago

We are now in a warm period following the end of the last glacial period colloquially known as an ice age. This glacial period peaked about 18,000 years ago, and by 12,500 years ago, our current inter-glacial warming was established. It ushered in a change in human life style that is probably the most revolutionary we have ever experienced.

It is startling to realize that *Homo sapiens* have been hunter-gatherers for several hundred thousand years, and that settled farming began just over ten thousand years ago. Farming is what made permanent cities possible, and it is out of these cities that civilization grew and spread into encompassing empires.

Early civilizations also supported some of our great inventions – the wheel, metal work, ceramics, domesticating animals, writing, the idea of democracy, and the great religions. Unfortunately, it also intensified war.

14. 🌱 10,000 thousand years ago:

The interglacial warming that started about this time supported the start of settled farming that in turn supported the first civilizations. Until then and for several hundred thousand years, we had been nomadic hunter-gatherers.

15. ═ 2,700 years ago:

As the settled living of civilization brought the leisure time required to think beyond the basic needs of the day, ideas of equality, democracy and the great religions of Buddhism, Confucianism, and the roots of Judea-Christian thought emerged.

16. ✝ 33 AD:

Christianity as a religion distinct from Judaism began.

Chapter 14

Farming Makes Civilizations Possible

13,000 to 2,800 Years Ago

Farming, Cities & Civilizations: *13,000-600 Years Ago*

Since the Last Freeze

The glacial period of the current ice age reached its peak 18,000 years ago. With the warming weather of our present interglacial, a dramatic change in the way people had lived for hundreds of thousands of years began. From this point on, the effects of humans on the very fabric of Earth's entire biosphere become unmistakable.

About 11,000 years ago, people began to cultivate crops in fields of fertile lands east of the Mediterranean. Supported by the reliable food supply this produced, sedentary farming communities gradually replaced nomadic life styles. This revolution in food production led to the world's first population explosion, the first cities, and to the differentiated and hierarchical societies of modern civilizations. Metal-work, writing and other technological inventions improved many lives. But with civilization also came slavery, inequality, and increasingly lethal warfare.

Today's Warm Period

13,000 years ago

About 25,000 years ago, the northern hemisphere entered a glacial period, or what is often called an ice age, lasting over 10,000 years. It peaked about 18,000 years ago. Then the climate began to get warmer, sometimes erratically, and our present interglacial began about 13,000 years ago. Although humans have lived through many warm periods, the response this time has been unlike anything that ever happened before. The most recent epoch that eventually brings us to our life styles of today had begun.

At first people lived as they had earlier, a nomadic life foraging and hunting. Then they deliberately began to cultivate grain. Although it might have seemed a small thing then, this change revolutionized human life.

Holocene Epoch   13,000 ya – present

Farming, Cities & Civilizations: *13,000-600 Years Ago*

All recorded human history has taken place during the present warm period. By 11,000 years ago, the ice covering much of North America, Europe and Siberia had melted. Deserts shrank, forests expanded, tropical forests returned. The effect on both plants and animals was pronounced. Changes in human life styles led to a transformation of ultimately global proportions.

With rising water levels caused by interglacial warming, land bridges disappeared and created four distinct populations. The world's human inhabitants were separated into four isolated landmasses for 9,000 years:

- Africa, Europe and Asia formed the "Old World"

- The Americas made up the "New World"

- Australia and Papua New Guinea created a third zone

- The Pacific Islands formed a fourth zone 4,000 years ago

Life developed in each zone independently. How it proceeded depended on climate, indigenous plants and animals, the success or failure of the foraging and nomadic lifestyles, and the potential for adopting customs from neighboring peoples. It is only recently that we have begun to understand the causes of these fascinating and arresting differences.

Chapter 14: Farming Makes Civilizations Possible - *Since the Last Freeze*

Beginning of Farming

11,000 years ago

Farming marks the beginning of what is called the Neolithic Revolution or New Stone Age. As big game was becoming harder to find, people in Asia began to supplement hunting by deliberately planting indigenous wild grains for harvesting. Over the following millennia, the number of domesticated plants increased and suitable animals were also farmed.

Farming brought new diseases and communities were more vulnerable to famine when crops failed. So early farmers were not as healthy and did not live as long as peoples in foraging or nomadic societies. Yet, by 3,500 years ago, farming had replaced foraging in large parts of Asia and Europe and today hunter-gatherers survive only in a few places usually not suitable for planting. Ultimately, agriculture led to the transformations upon which modern history rests.

Holocene Epoch   13,000 ya – present

Farming, Cities & Civilizations: *13,000-600 Years Ago*

more about...
the agricultural revolution

Farming revolutionized how we get our food and what we eat, where we live, and even how we fight.

- Farming changed our food supplies. Legumes and cereals like barley, wheat, rice and corn were farmed first because they grow fast and are easy to store. By Roman times 2,000 years ago, almost all our modern crops were cultivated. Today, domesticated crops account for 80% of all our food.

- Permanent settlements around farmlands replaced nomadic camps, and settlements grew into towns and cities. Today over half the world's population live in urban areas, fed by the farmers who ship their food to them.

- Farming intensified warfare. Lush fields were attractive not only to the farming communities which planted the crops, but to migrating tribes who saw them as rich sources of food for themselves. Warfare between nomadic tribes had traditionally been limited by the option of simply finding another field to forage if the price of war was deemed too high. For the sedentary community, war had higher stakes. If they lost their fields, the choice was between starvation and slavery.

To protect their lands, farming communities resorted to more organized warfare. They moved their homes closer together, often near water, and onto elevated positions where they could hold the high ground against invaders. Walls built for security gave the city specific demarcations and its inhabitants a clear identity.

Chapter 14: Farming Makes Civilizations Possible -
Since the Last Freeze

First Cities

10,000 years ago

The earliest cities like Jericho in Lebanon, Jarmo in northern Iraq, and Catahuyuk in Turkey, were built in the shadow of mountains where rain was plentiful. Whether any of these settlements then would be called cities by today's standards is debatable. They never had more than ten thousand people and usually less than a thousand. But these fixed communities show an unmistakable alternative to nomadic life.

Critically, farming provided enough food to support specialists, craftsmen, soldiers, even thinkers and teachers, and all sorts of things useful for a sedentary life appeared – pottery, bricks, baskets, and linen textiles. Supported by their farming hinterlands, cities became powerful centers for trade and innovation.

Holocene Epoch   13,000 ya – present

Farming, Cities & Civilizations: *13,000-600 Years Ago*

more about...
city life

City life was quite different from the wandering life of the nomadic tribe. Despite huge advantages, cities also generated new problems of their own:

• Food had to be brought to the centers of population, and waste had to be removed. The challenges of defending the increasing riches of the city also mounted as the standard of living increased.

• The complexity of city life decreased the egalitarian structures of nomadic groups, replacing them with a hierarchical structure under a king or priest leader. Different contributions to the common good were given different social status and specializations that did so much to improve life in the city also helped to generate social inequality. City life often benefited those with access to its privileges more than it did those further down the social ladder.

• Discrimination frequently fell disproportionately on women. In nomadic tribes women and men usually shared equal status. In sedentary communities, men more often occupied the positions of highest prestige, leaving women with less opportunity to achieve recognition in their own right. There were women in positions of leadership and influence. But in many patriarchal societies, women's activities outside the family were increasingly circumscribed and subordinate.

Chapter 14: Farming Makes Civilizations Possible -
Since the Last Freeze

Metal-Working

9,000 years ago

One of the most revolutionary processes humans ever devised began when they started to hammer copper into tools. Metal sickles and ploughs pulled by domesticated animals produced bigger and better crops that fed more people. Metal saws and axes felled trees more efficiently than cutters crafted from stone, and made serious carpentry possible. Metal was used to craft ships and chariots, which in turn accelerated trade and speeded up the transport of people, goods, and ideas. Metal weapons also made war more lethal and eventually shifted the balance of power to those who possessed them.

Innovations in metallurgy have never stopped. Looking at how much of our modern technology still depends on new metal creations, one might conclude that metallurgy quite possibly changed the world as much as farming.

Holocene Epoch   13,000 ya – present

Farming, Cities & Civilizations: *13,000-600 Years Ago*

more about…
the stages of metal

The use of metal to replace stone tools is divided into three stages that occurred in different times throughout the world. In some societies metallurgy never developed and technically they are still in the Neolithic or new stone age.

- The Copper Age: Copper implements were hammered at least 9,000 years ago and were used alongside stone tools. The oldest tools have been found in copper-rich Anatolia, today's modern Turkey.

- The Bronze Age: Bronze tools were made at least 7,600 years ago in Thailand and then in the Middle East. Bronze is much harder than copper because it is an alloy of copper and tin and making it requires the skill to keep a fire hot enough to fuse the metals together.

- The Iron Age: Iron implements first appeared mainly as weapons in the Middle East about 4,500 years ago. Iron ore is plentiful and can be hammered into shape but it is much more versatile if it can be moulded. Possibly because moulding or casting iron requires very high temperatures, iron implements were not widely used until 3,000 years ago.

Chapter 14: Farming Makes Civilizations Possible -
Since the Last Freeze

High Civilizations

6,000 years ago

"High civilization," is a vague term describing city-states during a great creative era that lasted for close to three thousand years. The oldest found so far, the cradle of civilization, is Sumer in Mesopotamia, now Iraq. The ideas and inventions that began here greatly influenced other high civilizations and still imbue Western and Islamic philosophy, culture, religious belief, and law.

People were living in Mesopotamia's Fertile Crescent between the Tigris and Euphrates rivers at least 10,000 years ago. The Sumerians arrived about 7.000 years ago, and within a millennium had created a modern civilization. They were inventive, pragmatic and open-minded, a confidant combination of tolerance and creativity. They built irrigation canals and wind-powered ships, fermented grapes to make wine, and, in a felicitous coincidence, the first glass. They used money, copper and bronze tools, designed the wheel to make carts easier to pull and fields easier to plough, and developed a system for writing. Their skilled bureaucrats oversaw their sea trade, arbitrated political disputes and international treaties, and thought up the idea of taxation.

Holocene Epoch   13,000 ya – present

Farming, Cities & Civilizations: *13,000-600 Years Ago*

more about...
mesopotamia's legacies

Through Greek, Roman and Judeo-Christian traditions, Mesopotamia's legacies may have influenced Western even more than Eastern cultures. Out of the Mesopotamia of this time came

- mathematical tables, algebra, and geometry

- a circle divided into six parts and an hour of 60 minutes

- an understanding of the seasonal equinoxes, the regular phases of the moon, and Zodiac names

- precursors of the Greeks' Aesop's Fables, and Plato's Dialogues

- the first records of a coherent musical system with a musical scale and intervals

- myths foreshadowing Biblical stories about creation, an earlier paradise, a great flood, the Cain and Abel rivalry, the Babel of Tongues, stories of a virgin birth, of death followed by disappearance for three days and eventual ascension to heaven, and apocalyptic visions of the future

- a rite of Baptism, rituals casting out devils, the concept of hell, and belief in the necessity of atonement

- written laws set down to protect the rights of the individual and the oppressed. Hammurabi's Law was meant to apply equally to everyone and influenced Greek, Roman, Islamic and Talmudic thought. Even the law dealing with mortgages goes back to Mesopotamia

Sumer's first leaders were democratically elected, but kings transformed their rule into a divinely-sanctioned hereditary monarchy, a capital idea from the point of view of kings that was not discarded in Britain and Europe until well into the last millennium, and still underpins rulers in parts of the world.

Chapter 14: Farming Makes Civilizations Possible - *Since the Last Freeze*

Writing

5,300 years ago

Art is probably as old as human beings. But writing did not develop until five thousand years ago. Of all the inventions of early civilizations, writing is among the most outstanding. Admittedly five thousand years ago, early writing was a cumbersome business, useful mainly to leaders and their scribes who understood the esoteric pictographs and hieroglyphics.

A great breakthrough came about 2,400 years ago when the Phoenicians devised the alphabet, the ancestor of the one with which these words are written and appropriately called "phonetic." The idea has been tinkered with over the centuries, but the fundamental system in which letters represent individual sounds rather than whole concepts is more flexible and easier to learn than any other system, and has never been replaced. It was the biggest advance in mass communications since talking.

Holocene Epoch   13,000 ya – present

Farming, Cities & Civilizations: *13,000-600 Years Ago*

more about...
early civilizations

Writing isn't our only legacy from the first high civilizations.

• In Egypt, city-states were developing alongside the Nile by 7,000 years ago. Civilization reached its initial zenith under the first dynasty of the Pharaohs between 4,664 and 4,158 years ago. They have given us a calendar year with 365 days and 12 months.

• In India, the Harappan civilization was flourishing on the Indus River by 3,500 years ago. Living in cities as big as 30,000, people fired bricks strong enough to build canals and docks, used a sophisticated system of writing, weights and measures and invented the mathematical concept of zero.

• The Shang and Chou Dynasties on the Yellow River in northern China sprang from the first settlements begun around 7,000 years ago. The first high civilizations between 3,700 and 2,256 years ago were ruled by a literate monarchy, and laid the foundations for many permanent characteristics of Chinese society. They left a heritage of silk, exquisite jade, stone carvings and bronzes.

Many people flourished in these high civilizations. Yet larger populations and improved weapons once more made war more aggressive, and the slaughter or slavery of men, women and children of vanquished enemies was the accepted norm. "Civilized" life was more brutal for some than in "primitive" societies.

Chapter 14: Farming Makes Civilizations Possible - *Since the Last Freeze*

The Wheel

4,000 years ago

The Sumerians used the wheel on farm carts but it was a cumbersome construction and the full potential of the wheel was not immediately evident. Then as the climate began to cool, nomads started to leave the steppes in search of more fertile fields. They brought with them a terrifying innovation - horses pulling wheeled chariots carrying men armed with bows and arrows. They moved into the Middle East, Europe, India, and China, wielding their fearsome weaponry. Old ways and new mixed in dynamic, if uncomfortable, exchange before the old orders gave way to new civilizations.

Most of the advances of earlier civilizations were not lost. The new lords recognized the benefits of much that was already there and made it part of their own regimes.

Holocene Epoch   13,000 ya – present

Farming, Cities & Civilizations: *13,000-600 Years Ago*

Peoples like the Eblaites, Elamites, Hurrians, and Hittites, Amonites, Assyrians, Phyrgians, and Phoenicians, Hyksos, Dorians, Hebrews and Persians added to a creative upheaval all over Eurasia. Kingdoms, empires, rulers and dynasties replaced each other for a millennium and a half before influence shifted from the Middle East to the Mediterranean.

- A major catastrophe, possibly a flood, devastated Sumerian cities 4,800 years ago. They were conquered by Sargon of Agade's Akkadian Empire which itself was replaced by the Babylonian Empire six centuries later. The Assyrians and Phyrgians moved into the north, and in what is now part of Lebanon, the Phoenicians took over Canaan and its old city of Jericho.

- Egypt was conquered by the Hyksos who set up Egypt's second Dynasty.

- In Crete, the Minoan civilization that had thrived on shipping for more than a thousand years was devastated by a massive volcanic explosion and never recovered. It was taken over by the Greek Mycenaean and finally destroyed 3,100 years ago by invading Dorians.

- In India, the Harappan civilization collapsed 3,750 years ago. We don't know if it was the result of foreign invasions or environmental catastrophe.

- In Western Europe, tribes, possibly Celts and Gauls, built the stone henges and circles that still dot the landscape from Scotland to Brittany. They reflect advanced astrological and mathematical calculations but their purpose remains a mystery.

Chapter 14: Farming Makes Civilizations Possible -
Since the Last Freeze

The Israelites

3,800 years ago

Among the semi-nomadic tribes of Semitic peoples during this time were the Israelites. According to their tradition preserved in Hebrew scriptures, Abraham led his tribe, selected by God to be a great nation, from the city of Ur in what is today Iraq to the Chosen Land of Israel.

Two more worldwide religions later also came to believe in the same God as Abraham's people. The tragic paradox is that today, Jews, Christians, and Muslims all believe they are uniquely God's chosen people, and that God has promised Jerusalem and the Chosen Land to each of them alone.

Holocene Epoch   13,000 ya–present

Farming, Cities & Civilizations: *13,000-600 Years Ago*

more about…
the god of the hebrews

The Israelites or Hebrews were the first people to arrive at a full-fledged permanent monotheism, the belief that there is only one god. The Hebrew God, Yahweh, transcended place and moved with the Hebrew tribes as they moved, making their god a particularly useful protector against threats from those whose gods reigned over only their patch of land.

The Israelites had great leaders and kings like Saul, David, and Solomon. Yet, perhaps their most enduring bequest came from their prophets who preached that God was the highest moral authority to which even kings were subject.

This was a significant intellectual leap. A king could conquer the world but his duty to God could not be abrogated. Two further world religions, Christianity and Islam, also preached that there are ethical standards to which leaders must submit. Even leaders in a democracy, the form of government about to be invented in near-by Greece, are, theoretically at least, to be held accountable, if not to God, then to the people. It is a concept that underpins the trials of war criminals today who may not have disobeyed the laws of their country but nonetheless are judged guilty of crimes against humanity.

Chapter 14: Farming Makes Civilizations Possible -
Since the Last Freeze

People Reach Oceania

3,500 years ago

The antipode of Africa is where you would arrive if you started to dig in Africa and didn't stop until you came up on the other side of the world. It's unlikely to be worth the effort because you would come up in the middle of the Pacific Ocean stretching roughly from the north to south poles and half way around the equator. In the middle of this vast watery expanse are flung the Pacific Islands or Oceania, with alluring names like Hawaii, Tonga, Polynesia, Easter Island, Micronesia, Fiji, and Samoa.

People from Taiwan or South China were probably the first to reach Oceania – or at least to survive there – between 4.000 and 3,500 years ago. It was the fourth and last region of the world to be colonized by *Homo sapiens*.

Holocene Epoch   13,000 ya – present

Farming, Cities & Civilizations: *13,000-600 Years Ago*

Despite redoubtable boat-building and navigational skills required to reach even the nearest Pacific Islands in the first place, booking a reliable round trip to an Oceanic island to anywhere else was not possible until the 17th century and even then it required something more arduous than a call to the local travel agent. So once the original migrants arrived on a Pacific Island, their isolated communities evolved different cultures in varying island conditions.

• In areas where farming is not feasible, Oceanic societies have remained small hunting-foraging tribes even today. Their lives may appear simple, but Westerners cannot survive there without their help. Their societies are egalitarian, and have an informal system of internal conflict resolution. It's more like the "I'll have a word with him" than "I'm going to report this to the police" approach.

• Where populations grew in density, a division of labor developed with a centralized, often hereditary, hierarchy. Conflicts increased and warfare intensified. Larger islands like Hawaii and Tonga developed proto-states with a strict class system and populations burgeoning to 50,000. With success – or at least with size - came literacy, slavery, an elite life style for those at the top, laws, and judges.

• Sedentary societies often constructed monuments like the statues on Easter Island, Oceanic versions of the pyramids or today's skyscrapers.

• Marriage arrangements, sexual partnerships, gender roles, child-rearing practices, and parenting responsibilities differed among Oceanic cultures, suggesting that what many Westerners had come to think of as "natural," was probably more "cultural." than they thought.

Chapter 14: Farming Makes Civilizations Possible -
Since the Last Freeze

Civilizations in the Americas

2,800 years ago

Although people may have reached central America as long as 40,000 years ago, New World civilizations seem to have been slower to develop than in the Old World. There were fewer crops and less suitable land for domestication, and variations in climate bands and mountain ranges limited the exchange of ideas and inventions that so enriched the Old World.

By 2,800 years ago, however, early civilizations in the New World were broadly similar to those in Eurasia several millennia before. Where farming emerged, sedentary groups created permanent settlements with widespread trade, political hierarchies and leaders whose authority was absolute and often cruel. Armies quelled rebellions and gruesome human sacrifices were offered to propitiate capricious gods. As in the Old World, astronomy, mathematics, architecture, engineering, religious thought, art and technological skills advanced. Perhaps because of the terrain, the wheel was not used until it arrived with the Europeans in the 17th century AD.

Holocene Epoch   13,000 ya – present

Farming, Cities & *Civilizations*: *13,000-600 Years Ago*

The Olmecs are the earliest known civilization in America who spread from eastern Mexico across most of Central America by 2,800 years ago. The Olmecs planned their cities and wrote with hieroglyphics but like the Chavin culture in Peru 2,400 years ago, they disappeared. Later New World civilizations emerged in different geographical areas.

The Mayans occupied the rainforests of Central America 3,000 years ago and reached the height of their power between 600 and 900. They had an alphabetical and pictorial script, possessed advanced mathematical and astronomical knowledge and knew that time extended at least hundreds of thousands of years into the past – a good deal longer than the 4,000 years estimated by 19th century European scientists. The Mayan language is still spoken by at least two million people.

The Incas in the Andes Mountains of Peru began to amass their huge empire about 1100. They worked gold, silver and bronze, their weaving was unsurpassed, and their doctors and surgeons were among the most skilled in the world. They were brilliant engineers, building mountain roads, bridges, tunnels and monuments with perfectly fitting blocks without mortar. Aqueducts ferried water in and out of their cities, and terraces turned steep mountainsides into fertile farms. The Incan language is used today by Native Americans throughout the Andes.

The Aztecs believed they had descended from the Olmecs. A nomadic people, they settled in what is now Mexico City in 1200 shortly after the Incan empire began. They eventually built a highly stratified empire with expansive trade and tribute. Their religious rituals were directed toward supplicating the gods they believed controlled the natural forces they knew they depended on.

Chapter 14: Farming Makes Civilizations Possible - *Since the Last Freeze*

Chapter 15

Democracy and the Great Religions

2,700 years ago to 1333

Farming, Cities & Civilizations: *13,000-600 Years Ago*

Glimpses of Equality

As the egalitarian traditions of nomadic life were replaced by the hierarchical differentiation of early civilizations, the belief in human equality was greatly diminished. It re-emerged in faltering steps as new civilizations grew up around the Mediterranean. 2,700 years ago, the Greeks of Ancient Athens set up a system of government called democracy in which the rulers were subject to the will of its citizens. Rome subsequently granted citizenship to all its free inhabitants, and Christianity came up with the radical idea that all Christians are equal before God.

Non-Christians, however, were not necessarily equal and at the beginning of the second millennium, the pope sanctified the crusades against unbelievers. As many adherents of the new religion, Islam, also believed it was a sacred duty to kill infidels, Eurasia was torn apart by centuries of war justified by religious belief. Despite the military defeat of the crusaders, medieval Europe became a prosperous and Christian group of nation states. Calamities of the 14th century nonetheless signalled a radical change in the established order.

Greek Democracy

2,700 years ago

The center of civilization shifted to the Mediterranean with the ascendancy of the Greeks. In the city of Athens, they set up the Polis, a city-state whose rulers were subject to the will of its citizens. Although citizens included only men and not women, foreigners, or slaves, the democratic principle that rulers should not have unbridled and arbitrary power but should serve the people was put into limited practice. Today it is honored, at least in theory, throughout much of the world.

Holocene Epoch   13,000 ya – present

Farming, Cities & Civilizations: *13,000-600 Years Ago*

more about...
the greek miracle

Historians call this period "the Greek Miracle." Democracy is its most well-known achievement, but its vibrancy and creativity led to much else that we still prize.

● In philosophy, people such as Socrates, Plato, and Aristotle explored the world with rational deliberate inquiry.

● Hippocrates, the father of medicine, gave us the Hippocratic Oath.

● Pythagoras and others gave us the foundations of geometry.

● Democritus suggested that all matter was made up of atoms; others later debated whether the sun goes around the Earth or the Earth around the sun.

● Homer wrote the Iliad and Odyssey, Herodotus wrote the first world history, and the first theatres appeared.

● The transcending standards of excellence of Greek sculpture and architecture still endure.

The Greek Miracle was brought to an end by the Peloponnesian Wars, a series of debilitating hostilities mostly among the Greek city-states themselves. City-states collapsed when another Greek, Philip of Macedon and his son, Alexander the Great absorbed them into a vast but short lived empire which spread Greek ideas far beyond the country of their birth.

Chapter 15: Democracy and the Great Religions -
Glimpses of Equality

Buddha and Confucius

Siddhartha Gautama, or Buddha, was born a prince of the warrior caste in India about 500 years before Christ. As a young man, he abandoned his comfortable life for asceticism and meditation, preaching that life's ultimate goal was to achieve ever higher states of consciousness. He believed caste and social status were irrelevant. Instead, through disciplined renunciation and self-abnegation, a person could escape the recurring cycle of rebirth and reach Nirvana, a state of cosmic unity and self-obliteration in which the individual essentially no longer exists. Buddhism became the first religion to spread beyond its original borders and today is one of the world's major religions.

Buddha, K'ung-fu-tzu, more familiarly known in the West as Confucius, was born in China about the same time as Buddha. In despair at the state of public service, Confucius turned to teaching. He believed above all in the established order and taught that the family, government, and other social institutions should be honored and preserved through personal integrity and disinterested dedication. The influence of Confucian principles in maintaining traditional observance and appropriate decorum is still evident in Chinese culture.

Holocene Epoch   13,000 ya – present

Farming, Cities & Civilizations: *13,000-600 Years Ago*

more about...
religion in india and china

- In India, Hinduism evolved over thousands of years, encompassing the worship of early gods like Shiva and Vishnu, and Vedic thought that arrived with the fall of the Harappan civilization. Hindus believe in the unity of all things, an idea that is almost impossible for individualistic and atomistic western thought to comprehend fully.

The caste system was integrated into the Hindu doctrine, teaching that one must carry out the duties of the caste into which one was born in order to be reborn in a higher form in the next transmigration. The castes, which became absolute and rigid, were Warriors, priests or Brahmans, common people, and what became the Untouchables, who were considered so unclean that even their shadow cast on food polluted it.

- China's earliest religions were directed toward gods of nature, whose desires were often divined by oracles. Unlike Western philosophers with their anguished search for meaning beyond the present, most Chinese thinkers, including Confucius, have been relatively unconcerned about a final judgement or life after death. Their philosophies have been primarily pragmatic and ethical, concentrating on maintaining social order and harmony.

- At the same time as Confucius, the teacher, Lao-Tse, was preaching his creed of Taoism. Lao-Tse urged his followers to adopt a life of simplicity and detachment from material luxury in submission to the Tao, or Way, the cosmic principle that sustains the order of the universe. Taoism, like Confucianism, did not ferment social change.

Chapter 15: Democracy and the Great Religions -
Glimpses of Equality

Roman Empire

2,027 years ago

Rome, the City of Seven Hills on the River Tiber, was occupied by shepherds around 2,800 years ago. It was a powerful republic within three hundred years and by 2,027 years ago, Caesar Augustus became the first emperor of a burly, pragmatic, and often civilizing empire. At its height the Roman empire covered most of Europe and North Africa. Its military might and engineering feats – roads, bridges, plumbing and architecture – improved the lives of many.

The Romans were gifted administrators, surprisingly tolerant of any local customs and beliefs that did not challenge Roman authority. The empire created *Pax Romana*, the Roman peace that enabled different, even opposing, peoples to live together in prosperity for centuries. In 212 AD, all free inhabitants of the empire were granted citizenship.

Holocene Epoch   13,000 ya – present

Farming, Cities & Civilizations: *13,000-600 Years Ago*

more about…
the romans

Roman rule lasted almost a thousand years. Its legacy still survives even beyond the borders of its empire:

- Except in England, the codification of Roman law underlies the structure of modern legal systems in Europe, in Latin America and in parts of the United States and Africa.

- Roman roads, forts, bridges, and cities still shape the map of much of Europe.

- Roman Latin is the basis of today's Romance languages.

- Our calendar of 365 days, although Egyptian in origin, was promulgated by Julius Caesar throughout the empire.

- Our day of rest each week started with the Jewish Sabbath, decreed by Caesar for everyone.

- The Silk Road, opening up trade between China and the rest of Europe, began during the Han Dynasty whose reign lasted in China from 200 BC to 200 AD.

Perhaps most outstanding of all, universal citizenship and another *Pax Romana* remains today the elusive goal of European politicians.

Chapter 15: Democracy and the Great Religions -
Glimpses of Equality

Chapter 16

Christianity

33 AD to 1333 AD

Farming, Cities & Civilizations: *13,000-600 Years Ago*

Religious and Political Power

Christianity began in Israel among a small group of Jews who believed that Jesus was their promised messiah. Christians were persecuted by the Roman authorities until the conversion of Constantine, when Christianity began to develop not only as a religious but also a significant political power.

With the Black Death the political authority of the Roman Catholic Church began to weaken. Conflicts of one kind or other between religious and political authority, however, have continued to this day.

Christianity's Beginnings

33 AD

According to Christian tradition, Jesus of Nazareth was born in Judea. He preached a gospel of love, and his followers believed he was their promised savior. Perceived as a threat by the Jewish religious leaders, he was accused of blasphemy and brought before the Roman governor, Pontius Pilate, who agreed to Jesus' crucifixion in the year 33. After his death, Jesus' followers believed he rose from the dead, and that those who followed his teachings would join him in eternal life.

Christians were persecuted, sometimes brutally, by the Romans for several centuries. Nonetheless, under protection of Rome's Pax Romana, Christianity spread and when Constantine became the Roman emperor in 324, he extended imperial favor to their religious practices. Subsequently, Rome recognized the value of co-opting religious beliefs to control the populations and support its political power. Eventually Christianity itself replaced the Roman Empire as a major political power throughout Europe.

Today, many Christian churches maintain Roman practices. Instead of the practice of early Christians in which priests were appointed by the communities which they served, leaders such as cardinals and bishops are appointed from above, bishops reside in palaces, and the use of ceremonial robes and forms of deference are adapted from traditions first belonging to the Roman Empire.

Holocene Epoch   13,000 ya – present

Farming, Cities & Civilizations: *13,000-600 Years Ago*

more about…
christianity

Early Christianity expanded the incipient concept of equality inherent in the Greek ideal of democracy. No longer was it justified to discriminate against other Christians because they were Jew or Greek, slave or free, male or female. All peoples were equal before God and social status, wealth, ethnic background or sex did not diminish or enhance their value. It is a principle that has occasioned heroic acts and fierce defense of the downtrodden through the centuries.

The doctrine that prestige and wealth are of trifling importance before God was more revolutionary than it may sound. It was to take another 1,500 years before Christian churches agreed that peoples of the world colonized by Europeans were also human, that slavery was inherently wrong, and that the fundamental rights of all people, whatever their religious persuasion, were equal and inviolable.

Chapter 16: Christianity - *Religious and Political Power*

Barbarians Sack Rome

410 AD

Maintaining the borders of a huge empire and meeting the needs of its population finally sapped Rome's strength. Weakened by internal power struggles, and assaulted by "barbarians" like the Goths, Vandals and Visigoths, Rome itself was repeatedly sacked. In 410, the Visigoths broke through the city defences and the members of the Senate took their own lives. After almost 1,000 years, Rome's hegemony collapsed.

With the loss of Roman authority, nomadic invaders swept across Western Europe, and people there sought new sources of security. Those tied to the land looked for protection from landowners who often titled themselves lords and even kings. With knowledge kept alive in the Christian monasteries, Christianity was gradually acknowledged as a major civilizing force and papal authority expanded. Over time, many of Europe's tribal invaders converted to Christianity, often a good option for pagan leaders whose gods periodically demanded the sacrifice of the leader himself. Christianity led to a more homogeneous society, blurring initial differences among the new immigrants.

Holocene Epoch   13,000 ya – present

Farming, Cities & Civilizations: *13,000-600 Years Ago*

more about…
the barbarians

"Barbarian" has become a pejorative term often applied to newcomers whose arrival may be fiercely resisted as foreign and uncivilized. But ten thousand years of history shows that new arrivals often bring new energy to stable cultures. Two groups illustrate how many of our own cultural landscapes are shaped by migrants who came so long ago we now think of them as our ancestors.

For many of us, they are.

- The Celts may have descended from Neolithic tribes in Eastern Europe. For centuries, they roamed – some say raged - throughout Europe. They were the Gauls who so harassed the Romans, who migrated into Asia, Spain, Italy, and who are probably ancestors to many Germanic peoples. They arrived in the British Isles between 4,000 and 3,200 years ago. When the Irish Celts converted to Christianity, they brought a mystical aura perhaps from early eastern religion, making the Celtic Church distinct from its Roman counterpart. The Celts, by then settled in the British Isles, were themselves among those attacked by another wave of outsiders, the Vikings.

- The Vikings, or Norsemen, from Scandinavia ransacked the British Isles, Europe and North Africa from the eighth to the eleventh centuries. They established themselves in Iceland, Greenland, and even Canada. In northern Russia, the Slavs' name for them, "Rus," was given to the whole country. They are remembered for their long boats and terrifying raids. But they also set up the "Althing," an incipient Parliament in Iceland. Slavery or even serfdom never took hold in northern England where they settled, and a fierce independence still survives as part of their legacy there.

Chapter 16: Christianity - *Religious and Political Power*

Byzantium

527 AD

Constantinople had been dedicated as one of the two capitals of the Roman Empire in 330 and when Rome fell, an emperor in Constantinople ruled over the eastern half of Christendom. It was called the "Roman Empire" for another 1,000 years, but all of Europe was never to be ruled from Rome again. After 527, when Emperor Justinian tried unsuccessfully to reunite East and West, Byzantium became a Greek, even oriental, empire with a distinctly eastern orientation.

Byzantium was more successful facing East than West. It fended off unwelcome overtures from a powerful Iranian Persian empire in 626, repelled centuries of Islamic attacks, and in the 12th century halted the advance of the Mongols, pagan tribes led by Genghis Khan. It wasn't until 1453 that Moslem Turks of the Ottoman Empire sacked Constantinople, renamed it Istanbul, and ended Byzantium's long rule.

Holocene Epoch   13,000 ya – present

Farming, Cities & Civilizations: *13,000-600 Years Ago*

After the collapse of Rome, cultural and religious differences between East and West began to be problematic.

- At first, differences were theological. In the West, Roman rule and its tolerance of religious diversity had left an incipient separation of church and state. Western secular and spiritual rulers often tried to dominate each other, but at least in theory, they separated their domains.

  The Eastern emperor, on the other hand, believed that as God's viceroy he himself held both spiritual and temporal power. The East did not accept the pre-eminence of the Bishop of Rome, nor did it tolerate heretics or Jews whom the Emperor decreed were enemies of Christianity and so of Byzantium.

- Conversions to Christianity among the invading tribes who had settled in their new lands also deepened the East-West split. Norsemen and Slavs in Russia and the Balkans converted to the Orthodox Christianity of Byzantium, while Slavs in Poland and most tribes in the West converted to Roman Christianity.

  When large numbers also converted to Islam, Europe was rent by three religious views, each possessing a compelling culture and all-encompassing creed. Their influences are still shaping attitudes and values in the modern world.

Chapter 16: Christianity - *Religious and Political Power*

Islam

622 AD

Muhammad was born in Mecca about 570, and grew up an orphan in a Bedouin tribe. He was a visionary, and in 622, his followers began to preserve his words in what is now the Koran, the sacred book of the religion of Islam. Its creed is one of submission to God's will, regular prayer, and the brotherhood of believers in a society in which, according to many believers, women should be subject to men. Allah, who is the same as the God of the Jews and Christians, is the only true God.

Islam developed one of the most sophisticated, scholarly and artistic cultures the world has known, preserving much of human knowledge as it spread across Eurasia. Today one-sixth of the world's population are Muslims. Islam continues to exert not only a religious but a significant cultural influence.

Holocene Epoch   13,000 ya – present

Farming, Cities & Civilizations: *13,000-600 Years Ago*

more about...
islam

Islamic armies defeated the Persian army in 642, conquered
North Africa in 670, and Spain in 714. By a thousand years
ago Islam was at the heart of a diverse civilization reaching
from the glories of Spain on the west to China on the east.
While Europe was in a backwater, the benefits of Islamic
vibrancy were reflected in a confidant tolerance of ethnic and
religious differences. Ideas from Greece, Persia, India, Arabia
and Africa were revitalized and expanded, and poetry,
philosophy, mathematics, science, medicine, and art
flourished.

One of the jewels in its crown was Cordoba, Andalusia in
Spain, where Muslim power had arrived with the Berbers.
Until the Moors, as they were called, were finally pushed out
of Spain, their sophisticated and cosmopolitan culture
flourished there for 700 years. Their gardens of Alhambra in
Southern Spain, with their flowing waters, their balance, and
exquisite flowers and trees are still there today, oases of
serenity and beauty. Islam political leadership eventually
rested with the Ottoman Turks whose empire lasted until
1922.

Chapter 16: Christianity - *Religious and Political Power*

Charlemagne's Reign

771 AD

For six centuries following the fall of Rome, western Europe consisted mostly of kingdoms, really tribal organizations of earlier invaders under the leadership of strong men. Over the centuries, different rulers united lands under their control and lost them again, but in 771, Charlemagne, a Frankish warrior, became king. He was to become a legend, the first ruler since the Romans to unite western Europe. He converted to Roman Christianity, and aggressively defended papal lands against barbarian attack. He was crowned Roman Emperor by the pope in 800.

Charlemagne was a greatly loved and charismatic leader, who valued learning and art, and his residence in Aachen became an intellectual focal point for the West. Although the unity of his kingdom did not survive long after his death, many today still look to him as the father of Europe.

Holocene Epoch 13,000 ya – present

Farming, Cities & Civilizations: *13,000-600 Years Ago*

more about...
the medieval church

Scholars are not sure how pleased Charlemagne was with his coronation and the implication that it was in the pope's remit to confirm his secular power. The papacy needed protection from invading tribes, and Christian warlords benefited from the legitimacy bestowed by Church approval. Yet this shifting alliance between church and state was always a wary one, each jealously guarding against the other's potential encroachment into its territory. They went forward in troubled lockstep, both kings and popes often blurring the lines of their separate authority.

The Church, and in particular the Roman papacy, had become an important force in medieval Western Europe, Christianizing many barbarian customs and festivals, and enforcing the principle of monogamous marriage. It became wealthy in its own right, ruling over its own lands and administering its own courts of justice.

By the end of the first millennium, Church doctrine permeated the very fabric of life, interpreting the meaning of every natural event and social interaction. It gave the year its rhythm of fasting and festivals, its days of worship and rest. It mediated the understanding of illness and health, of fortune and misfortune, of birth and marriage, of sin and punishment.

Chapter 16: Christianity - *Religious and Political Power*

The Crusades

1099 AD

The first crusade was ostensibly mounted by western Christians against the Moslem Saracens to regain the Holy Land from the "infidel," although some modern scholars suggest that political concerns may have been a more important influence. Crusaders initially recaptured Jerusalem but it was a temporary victory and new crusades continued for almost two centuries. During this time the Knights Templar, the Knights of St. John Hospitaller, and the Teutonic Knights became powerful and rich, eventually leaving a legacy of intrigue and secret wealth that continues to tantalize historians and conspiracy theorists alike. The last crusade ended in 1291 when Christian warriors withdrew in a final ignominious defeat.

Holocene Epoch   13,000 ya – present

Farming, Cities & Civilizations: *13,000-600 Years Ago*

more about...
the booty of war

The crusades had a major impact on Western Europe.

● Although Christians had raped, pillaged, burned, and murdered before, the justification for the crusades represented a significant theological step. Instead of practicing love to spread the Word of God, war became the Christian mission, even the passport to sainthood and heaven. The deeds of the crusaders were perceived by some as heroic and honorable but they were often vicious and out of control, leading even to sacking Christian Constantinople in an orgy of bloodlust.

● Yet, the consequences for the West were not chiefly the fruits of military disaster. The Arabs had built on the knowledge of earlier civilizations mainly lost to the West during the Dark Ages. Much of this knowledge was brought back home by the crusaders. Wind and water mills, glass, metal work, clocks, new techniques for ship building, gunpowder, paper, and the magnetic compass now began to enrich the West. Works of the great Eastern philosophers began to transform Western thought.

Chapter 16: Christianity - *Religious and Political Power*

Europe's High Middle Ages

1200 AD

There is a period of about 400 years at the beginning of the second millennium called the High Middle Ages when Europe was developing as a prosperous, energetic society. At the time, things may not have seemed to be changing much. The Church was the supreme spiritual authority, and magnificent churches and cathedrals rose as monuments of inspiration and outstanding skill in a glorious paean of faith. Monasteries, still almost the sole reservoirs of learning in a largely illiterate society, owned vast tracts of land, and exercised political influence at the highest levels.

Yet changes were occurring that, however unnoticed, would eventually undermine this settled order.

Holocene Epoch   13,000 ya – present

Farming, Cities & Civilizations: *13,000-600 Years Ago*

In 999, Muslim cultures were centers of science, art, medicine, and culture. Civilizations in China and India were more advanced than any in Europe. Few would have guessed then that Europe, filled with illiterate peasants, would exert an influence that in the next millennium would reach into every continent on the planet. How did it happen?

Universities were an invention of the Middle Ages. The first appeared in the 1100's in Paris, Bologna and Oxford. Salamanca in Spain and Heidelberg in Germany followed. They were still greatly influenced by religious values but higher education was moving beyond the monastery.

Franciscans and Dominicans were a new kind of religious order. They left the seclusion of the cloister, preaching instead as mendicant friars. Dominican mendicants epitomized in Thomas Aquinas were scholastics who brought the Greek thought of Aristotle to the center of Christian theology and paved the way for the objective study of the natural order.

The introduction of horses for ploughing crops, and water and wind mills for grinding grain created an agricultural revolution that massively increased the food people had to eat. Populations grew.

Gunpowder and explosives changed the way war was waged. Central authority had been ineffectual in subduing rebellious barons in their fortified castles but cannon fire penetrated the defences of thick walls and could defeat local landowners. As a result, power shifted, and the relative strength of kings and their nation-states grew. The uneasy balance between papal and civil authority was unsettled and was not to last indefinitely.

Chapter 16: Christianity - *Religious and Political Power*

Black Death

1333 AD

The bubonic plague or Black Death began in the Far East in 1333 and, carried by fleas and rats along trade routes, arrived in Europe in 1347. It was an excruciating illness that was almost always fatal within days. After five terrible years, 25 million people, one-third of the population in Europe, had died. There was similar devastation in the East. The disease returned in small deadly episodes for centuries.

The plague undermined two bulwarks of Western Europe's medieval society – land-based wealth and the authority of the Church. Labor shortages gave farm workers more bargaining power and many who had been tethered to the land moved to the cities for higher wages. Peasant revolts spread, class conflict increased, and the feudal system was destabilized. For some, religious observance intensified, reflecting manic, even bizarre, observance. But the plague had struck down the godly and ungodly alike, and others began to question the foundations of their faith. They no longer fully trusted the Church's teaching and their greater self-belief and independence began to weaken ecclesiastical authority.

Holocene Epoch   13,000 ya – present

Farming, Cities & Civilizations: *13,000-600 Years Ago*

more about…
the century of catastrophe

Besides the plague, other calamities in the 14th century helped fracture the social, economic, and political structures of medieval Europe.

The Great Famine: At the beginning of the century, the land was cultivated to its capacity to feed the growing population. Then in 1315, heavy rains flooded the fields and rotted the grain. When the floods reoccurred the next year, serious famine set in. Children were abandoned, work animals slaughtered, and some people may have resorted to cannibalism. In eight years, 15% of the population starved to death.

Jewish Massacres: Within a year of the plague's arrival in Europe, Jews were being blamed. Torture forced some to confess to poisoning the wells; for others even the pretence of a trial was unnecessary. Jews were burnt to death in Switzerland, Germany, France, Brussels and Spain. It did not slow the plague.

The Great Schism of the Papacy: Between 1378 and 1417, the authority of the Church of Rome was further weakened by two popes, each claiming simultaneously to represent the true Church. One set up court in Rome, another in Avignon, and proceeded to excommunicate each other and their followers. It was particularly difficult for the faithful who believed that to follow the "wrong" pope, even unintentionally, would result in eternal damnation.

The Hundred Years War: The war between France and England was fought sporadically mostly on French land from 1339 to 1449. Despite several heroic battles immortalized by Shakespeare, England was eventually defeated.

Chapter 16: Christianity - *Religious and Political Power*

Chunk VI

Modern Life Emerges

In the Last 600 Years

Changes that began with the Renaissance almost 600 years ago are so revolutionary that it is often hard to appreciate what a short time humans have lived like this. Modern life began with a wave of communications and transportation break-thrus which became the seed bed of the religious, scientific, industrial and political revolutions to follow. It was also the seed bed of centuries of warfare.

Waves of communication and other technological advances, accompanied by increased religious tensions, political upheaval, and warfare, continue to this day.

17. ➤ during the 1400's: The beginning of Europe's world-wide exploration and the printing press each introduced two <u>revolutions in communications</u> that transformed our understanding of the world and our place in it.

18. ✂ during the 1500's: <u>The Protestant Reformations</u> undermined the absolute authority of Rome and led to two centuries of religious warfare throughout Europe.

19. ✏ during the 1600's: The <u>scientific revolution</u> and Galileo's proof that the world orbited the sun led to serious conflict with Rome.

20. ✸ during the 1700's and 1800's: Advances in science and technology applied to production led to the <u>Industrial Age</u> transferring global economic dominance to Europe, and especially to Britain. The religious wars of the previous centuries were the impetus for <u>modern democracies</u> and the separation of church and state authority. The 1800's is marked by the height of European Empires and the American Civil War.

21. ✈ 20th Century: <u>Technological advances</u> we now take for granted began less than 100 years ago. Issues of <u>human rights,</u> and <u>conflicts of war</u> have proliferated while the space age and the internet have opened stunning possibilities that before could only be imagined.

Chapter 17

Revolutions in Communications

1400's

Modern Life Emerges: *In the Last 600 Years*

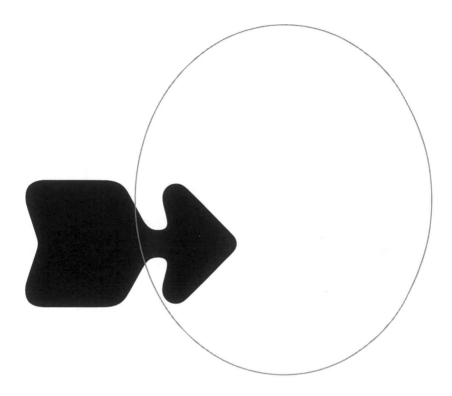

The World Becomes Global

The new beginning called the Renaissance began with two changes in communication that ultimately transformed human experience in the entire world. The first, in what we might today call long-distance travel, European ships discovered lands and people they had not known existed for ten thousand years. The second was the development of the printing press, which facilitated the spread of ideas beyond the confines of the monasteries to the common people.

Europe's Global Explorations

1434

The magnetic compass was invented in China, but European seamanship benefited from it the most. 600 years ago, sailors setting sail were the astronauts of their day. Nobody knew if there was a way to sail around the world or how big it was. A lot of people weren't even sure the world wasn't flat. Nobody knew what continents existed, or where they were.

The Portuguese were among the first Europeans to start the European sea-faring explorations of this age. In 1434, they pushed beyond the limits of their known world and sailed south along the west coast of Africa.

In 1492, Christopher Columbus discovered what today we call the Americas.

In 1498, Vasco de Gama reached India, opening an all-sea route for Portuguese trade with the East. In 1513, Portuguese ships reached China.

In 1522, Magellan's expedition returned from the first circumnavigation of the globe. The world was indisputably round.

In 1606, Europe learned about the presence of Australia and New Guinea.

This is when the world really started to became global.

Holocene Epoch   13,000 ya – present

Modern Life Emerges: *In the Last 600 Years*

more about…
Columbus and world exploration

Christopher Columbus is celebrated as the man who discovered America in 1492. But Columbus wasn't looking for America. He was looking for a route to the East by sailing west. He was right in his belief that Earth was round instead of flat, but underestimated its size. Instead of finding what he was looking for, he stumbled on Central America. When he landed in the Bahamas and Cuba and claimed them for Spain, he thought he was in India and so called the people there "Indians." Columbus, himself, never actually set foot on land that became the United States.

Yet the celebrations today may underestimate this remarkable voyage. Of course, the Americas had been "discovered" thousands of years earlier by the peoples who were already there when Columbus arrived. But the original voyagers had been unable to return to their origins and by the 15th century, nobody in Europe even knew about them. What Columbus found really was a new world for everyone in Europe.

Chapter 17: Revolutions in Communications -
The World Becomes Global

Printing Press and the Renaissance

1453

Just over 500 years ago, Johannes Gutenberg introduced to the West the biggest break-through in communications technology since the Phoenicians devised the alphabet. The first printing press would have been slow enough to bring any modern office to near a stand-still, but compared to the laborious hand copying that was all Europe had for thousands of years, its speed was revolutionary.

The first book printed with the new movable type was the Bible, which seemed benign enough. As printed material proliferated, however, its seditious potential for influencing public opinion and undermining the status quo gradually became apparent. The widespread and relatively inexpensive publication of books and pamphlets helped make Europeans more literate, better informed, and less gullible. The major institutions of Europe were about to be propelled into the modern age.

Holocene Epoch   13,000 ya – present

Modern Life Emerges: *In the Last 600 Years*

more about…
the renaissance

The printing press was invented during the Renaissance, a French word meaning "new birth," to describe a vibrant period during the 15th and 16th centuries. It eventually encompassed philosophical and theological thought, scientific discoveries, technological advances and the arts. It began in Italy where people were benefiting from a burgeoning mercantile trade.

Renaissance values mark a distinct change from medieval Europe. Although Renaissance people generally did not challenge the Church, Christian teaching was no longer venerated as the only source of legitimate knowledge. Glimmers of scientific thought were emerging and the classical writings of the early Greeks and Romans were studied with respect. The will of God became an insufficient explanation of events. Instead, natural causes, including man himself, were held responsible for what happens. Learning began to supplement piety as a meaningful human endeavor.

Giotto, Raphael, Michelangelo, and Leonardo da Vince, Dante, Erasmus, Machiavelli, and Donatello are among the Renaissance names whose genius we still recognize.

Chapter 17: Revolutions in Communications -
The World Becomes Global

Chapter 18

Religious Reformations

1500's – 1600's

Modern Life Emerges: *In the Last 600 Years*

Questioning Rome's Authority

The upheavals of the 14th century had weakened Rome's absolute authority, and with increased communications, education, and development of a merchant class, people in Europe developed new attitudes. They began to feel more individually responsible for what happened to them and the Roman Catholic Church was assaulted by the demands arising from the Protestant Reformation and its emphasis on the primacy of individual conscience and communion with God unmediated by Rome.

The ensuing power struggles led to almost two centuries of bloodshed throughout Europe and cost millions of lives.

Luther and the Protestant Reformation

1517

In 1517, Martin Luther, a scholarly young German monk, sent 95 theses to his superior, his bishop, and the pope in Rome and according to legend, posted a copy on the door of the Wittenberg church. Among other things, Luther was objecting to the selling of indulgences, essentially a money-raising scheme by Rome whereby people believed they were buying time off punishment in purgatory after death before proceeding to heaven. The printing press led to a rapid spread of Luther's ideas, tapping into a deep German resentment of the Italian papacy and clerical corruption.

Luther taught that salvation was not earned by works but justified by faith, and that the demands of individual conscience were more important than submission to the pope. Further undermining an all-encompassing role of the church, Luther preached that a private relationship with God could be guided by personal study of the Bible without church mediation. It was perceived as a radical view then and an attack on the authority of the Roman church that would have been unthinkable a century earlier.

Holocene Epoch   13,000 ya – present

Modern Life Emerges: *In the Last 600 Years*

more about...
the protestant reformation

Although it was not what Luther intended, the Protestant Reformation fractured the unity of Western Christendom.

• In 1555, the Peace of Augsburg agreed to a division of the German states on the principle of 'cuius regio eius religio' – each ruler would determine the religion of the state. With its emphasis on individual conscience, however, Protestantism itself began to fragment. The main Lutheran and Calvinist divisions were each splintered by doctrinal disputes, and sects proliferated as Protestantism spread through France, the Netherlands and Scotland. In England, Henry VIII, who earlier had defended the church against Luther, rejected the authority of Rome, closed the monasteries, and made the king head of the English church.

• Roman Catholicism responded with the counter-reformation. Lax morals and theology were censured and the Index set up to control what people read. The Inquisition, begun in the 13th century to suppress heresy, was reinvigorated, using terrifying powers to torture and execute heretics. Ignatius Loyola formed the Jesuits, who vowed obedience only to the pope in Rome and enforced his will with military discipline and a program of education.

Chapter 18: Religious Reformations -
Questioning Rome's Authority

Religious Wars in Europe

1524- 1648

The fragmentation of Christian doctrine created massive intolerance. Each group still believed that it alone held the true Faith, and pursued an often fanatical mission to eradicate error in others. Protestantism itself did not abandon the practice of coercing compliance through torture and death threats and Europe was torn by centuries of religious wars.

The fracture of the Roman Catholic monolith in Europe brought further criticism of the greed and corruption endemic in the church and exacerbated social and ethnic tensions. The wars, nominally called "religious," were often in pursuit of temporal power, land, and a social reform related not to one's ultimate place in heaven but to a demand for change in the present. A growing class of merchants, artisans, and bankers resented the church's intrusion into their secular affairs, the nobility resented the untaxed privileges of the clergy, and peasants rose up in revolt against their exploitation.

When the religious wars finally ended, secular rule was greatly strengthened, and philosophers were beginning to argue that it was not the role of the State to enforce a religious code of any persuasion. It was to be the seed of another revolutionary idea which, like the Cold War of the 20th century, avoided war at home for many years.

But the peace was not absolute. Europe would be ravaged again within two centuries by two world wars as devastating as the Wars of Religion.

Holocene Epoch   13,000 ya – present

Modern Life Emerges: *In the Last 600 Years*

The Wars of Religion reached across Europe and spanned almost two centuries. They were as often about power and wealth as they were about religion. But it was inevitably on religious grounds that opposing sides rallied.

● The German Peasant War, 1524-1525, arose at the time of increasing controversy begun by Luther, but it may have been driven by wealthy peasants defending their positions of pre-eminence.

● Warfare began between Catholic and Protestant cantons in Switzerland in 1531 and continued on and off until 1712.

● Dutch War of Independence from 1568-1648 continued for 80 years between the ruling Catholic Netherland Hapsburgs and Lutheran, Calvinist and Anabaptist Protestants.

● French Wars of Religion began in 1562 between French Catholics and Protestant Huguenots. Both sides received help from abroad and intermittent fighting continued as late as 1715.

● Wars in Scotland, England, and Ireland began with King Henry VIII's edict making him the head of the Church of England. Wars between 1639 and 1651 divided the population on religious grounds, beheaded a king, elevated the Protestant Cromwell, and produced martyrs among Anglicans, Catholics and Protestants.

● Thirty Years War, 1618-1648 was the last major overtly religious war in Europe. It began on German soil but ultimately involved all of Europe, and was one of the most destructive of modern times. Populations in some parts of Germany were reduced by 65% and European naval powers attacked each other in Africa, Asia, and Latin America. The Peace of Westphalia, which finally ended the fighting, laid the foundations of the sovereign nation-state we know today.

Chapter 18: Religious Reformations -
Questioning Rome's Authority

European Conquest and Colonization

1532

According to historical records, the Spanish conquistador, Pizzaro, conquered the Inca Empire with two hundred solders in 1532. He invited the Inca ruler, Atahualpa, to a meeting where he was kidnapped and his family murdered. Atahualpa agreed under this less-than-subtle pressure to convert to Catholicism, but Pizzaro, quite possibly with his eye on Incan gold more than on Incan salvation, had him executed.

As their ships reached the Americas, Asia, China, India, and European powers fought in mendacious competition. With amazing rapidity, they subjugated native peoples, pillaged their temples for gold and precious stone, exploited their natural resources, and brought missionaries to replace their cultures with Christianity. By the end of the 16th century, the Portuguese had claimed Brazil in South America and Goa in India. The Spanish had conquered vast lands in North, Central and South America, and the French, Dutch and English all had far flung trading centers and colonies. European colonizers were on a roll.

Holocene Epoch   13,000 ya – present

Modern Life Emerges: *In the Last 600 Years*

In less than a hundred years, European conquest brought about cultural destruction and a massive loss of life among native peoples of North, Central, and South America. How was this possible? Were the Europeans more advanced or intelligent? Were the native peoples more primitive, peaceful or naive? It's unlikely. The Europeans brought superior weaponry with them that they used with brutal arrogance. They also brought their lethal microbes.

- Many diseases that devastate humans begin with other animals. Measles, TB, small pox, influenza, cholera, the plague, typhus, polio, mumps, leprosy, and malaria were all transmitted to us from domestic cattle, pigs, ducks, dogs, cats, or birds. Through proximity to their domestic animals, Eurasians had endured recurring epidemics over the previous millennium, and people who had not died had developed immunities to them. The plague might have produced immunities protecting some Europeans from the Aids virus even today.

- Native Americans had no such immunities to diseases like the small pox conquistadores brought from the Old World. Within a century, a horrifying 95% of the original natives in the New World had died. Far more were victims of disease than of military defeat.

The colonizers had little respect for the peoples they invaded. But the Inca and Aztec Empires, and the tribes of the North American Indians did not collapse solely in the face of brutality, duplicity or superiority. They also collapsed in the face of European disease.

Chapter 18: Religious Reformations -
Questioning Rome's Authority

Chapter 19

The Scientific Revolution

1500's – 1600's

Modern Life Emerges: *In the Last 600 Years*

Rules of Evidence

The absolute authority of Roman Catholicism attacked by the Protestant Reformation was also undermined by the scientific revolution, which began with the Renaissance. Scientists themselves did not believe their studies interfered with the spiritual authority of the church. They most often believed, as do many scientists today, that the study of the natural world revealed new depths of an awesome divine mystery.

But conflict between church authorities and scientists broke out as early as the 13th century when Rome tried to silence those with whom it disagreed and forbid under pain of excommunication anyone to read Aristotle and other Greek thinkers. The philosopher/scientist Francis Bacon was imprisoned by the church for his defence of natural laws. Most famously, when Galileo offered proof of the Copernican theory that Earth revolved around the sun, he was threatened with the rack, and sentenced to house arrest for the rest of his life. It was a conflict not dissimilar from the controversy today between some believers and scientists about the theory of evolution.

Galileo and the Scientific Revolution

1632

In 1632, Galileo Galilei published his book suggesting that the Earth was not the center of the Universe but orbited the sun. A century earlier, Copernicus had said the same thing, but Galileo had empirical evidence to support his argument. For church authorities, this was dangerous and heretical. If the sun did not orbit the earth, where was God? where was heaven and what of the resurrection? Called before the Inquisition in Rome and put on trial, under threat of being stretched on the rack, Galileo, old and sick, recanted.

But the printing press had already widely disseminated his work and it was known that observation supported Galileo's, not Rome's, position. Scientists began to move out of Rome's reach, and the advance of science continued mainly in England and in other Protestant countries.

Holocene Epoch   13,000 ya – present

Modern Life Emerges: *In the Last 600 Years*

more about...
the scientific revolution

Although the recantation extracted from Galileo may be a watershed, the scientific revolution took place over several centuries.

- Careful observations of nature began during the Renaissance. Rules of perspective were laid out, human anatomy was studied, and map-makers strove to describe land and sea with less imagination and greater accuracy.

- Instruments for observation opened up worlds nobody had seen before. Telescopes, microscopes, barometers and better clocks made measurements more accurate.

- In 1620, Francis Bacon described the inductive method to discover the laws of nature. Valid scientific knowledge was acquired through systematic observation first, not by beginning with first principles using Aristotle's method.

- In 1637, Rene Descartes set out the principles of deductive reasoning applied to the scientific method.

- In 1687, Isaac Newton published *Principla Mathematic,* explaining the universe in purely mathematical and mechanistic terms. It stood unchallenged until Einstein's theory of relativity in the early 20th century.

The scientific approach replaced many religious explanations of the world. Conclusions were no longer accepted without question on the basis of faith but were tested by systematic observation. Most scientists remained committed Christians, and believed they were praising God by studying the glories of his creation. But the judgement of the Roman Church on questions of science was no longer universally regarded as valid.

Nonetheless, although few church authorities today would argue that the sun circles the earth or that angels keep the stars from falling to Earth, many still argue that religious belief can describe the natural world with infallibility and should overrule the findings of science.

Chapter 19: The Scientific Revolution - *Rules of Evidence*

Little Ice Age

1650

Although it's not often mentioned in history class, a Little Ice Age descended on the northern hemisphere between about 1350 and 1850, peaking between 1645 and 1715. It is "little" not because it confined itself to a conveniently small corner preferably near the north pole or the Sahara desert, but because for an ice age, it was short. But it was serious. And it was cold.

Agriculture was drastically reduced, and famines and disease killed millions of people. The population of Iceland alone fell by half and glaciers in the Swiss Alps buried entire villages. Plummeting temperatures froze rivers and canals in the Netherlands, the Thames in London, and in New York, people walked on ice from Manhattan to Staten Island.

Holocene Epoch   13,000 ya – present

Modern Life Emerges: *In the Last 600 Years*

more about...
ice ages and global warming

Ice Ages occur regularly on Earth about every 150 million years and last for tens of millions of years. Technically, Earth is still in an ice age that began several million years ago. Fortunately for us, however, Earth is in an "interglacial warm period" which began 13,000 years ago, and this Little Ice Age in recent centuries is a cold period that occurred in the middle of this warm period. If we can understand it, perhaps we can understand the future of our own climate a little better.

Several things may have caused the Little Ice Age:

• decreased sunspot activity that occurred during this time might reflect a fluctuation in the sun's energy output.

• increased volcanic activity threw up masses of ash. It may have blocked so much sunlight that an ice age set in.

• a reduction in greenhouse gases may have occurred as a result of decreased farming throughout Eurasia following the Black Death. This could have let more warmth escape, and cooled climate temperatures.

• once an ice age starts, it can perpetuate itself because snow and ice deflect sunlight, preventing it from re-warming the Earth.

Could it happen again?

• Although warm weather is the norm for Earth, scientists are concerned that the global warming now being engineered by the greenhouse gases produced by human activity may warm Earth to temperatures it has not seen for 4 million years. Ice would melt and raise water levels enough to put New York and London under water.

• Some climatologists studying ocean dynamics think that, paradoxically, global warming could cause another ice age. Atlantic currents are showing signs of radical shifts today. Should they stop funnelling the warm Gulf stream north, many people would face much colder, not warmer, weather.

Chapter 19: The Scientific Revolution - *Rules of Evidence*

Chapter 20

Industrial and Political Revolutions

1700's – 1800's

Modern Life Emerges: *In the Last 600 Years*

Utopias Re-Examined

By the 17th century, after two centuries of religious wars had wracked Europe, philosophers and politicians were conceiving a new social order. The rights of individuals were recognized in a political system called democracy, the role of the state no longer included the enforcement of religious practices, and industrialization was improving the daily lives of working people. For many, it was a vision of a new utopia. For some it arrived. For others it did not. For many, the new utopia was the Communism of Karl Marx.

The 20th century continued the transformation of people's lives through new technologies like the car and electricity. It also implemented the human rights of women throughout the Western world and continued to reverse some of the most blatant aspects of racism. But it also brought the world the brutalities of war on a scale never seen before.

Steam and the Industrial Age

1763

The industrial age began as power-driven machinery was invented to replace manual labor. In 1763, James Watt, a Scottish engineer, harnessed the energy of steam and put it to the service of manufacturing. The steam engine untethered mills from the stream-side, and factories moved closer to their sources of supply and commerce in the cities. When steam locomotives were used to power railway systems, land transport was revolutionized. Then the steamship did the same for shipping, freeing it from wind power's capricious unreliability.

In the end, industrialization quite possibly changed the daily lives of ordinary people as radically as the change from nomadic to sedentary life styles 10,000 years before.

Holocene Epoch   13,000 ya – present

Modern Life Emerges: *In the Last 600 Years*

more about…
early industrialization

In Britain, inventions sprouted in enthusiastic optimism. A generation of inventors built on the foundations of the scientific revolution tapped large coal deposits for fuel, and benefited from the colonial empire that provided raw materials and a market for finished products. It changed the landscape and the lives of almost everyone in fields as diverse as farming and pottery.

- Agriculture: The seed drill, crop rotation, selective breeding, and new root vegetables like potatoes increased food production and Britain's population.

- Pottery: In Staffordshire, waterwheels and windmills revolutionized production. As more people could afford to buy crockery, eating became more hygienic and safer.

- Textiles: The first textile factories opened in 1740 where the flying shuttle speeded up yarn-making. The spinning jenny, carding machine and printing roller greatly improved productivity, and when the entire system was integrated by the steam machine, the appetite for cotton ballooned. The cotton gin was designed in 1793 to clean raw cotton and the sewing machine was introduced in 1850. Until the industrial age, most people in England wore wool, often uncomfortable, damp, and infested. By 1800, factory production had made clothes cheaper and what people wore was vastly improved.

Machines could make things more cheaply and faster than making them by hand and as productivity soared, the lives of many ordinary people improved dramatically. Yet life was by no means universally better. Many traditional jobs and skills were lost, and within a century the squalor of factory workers led to social upheaval.

Chapter 20: Industrial and Political Revolutions - *Utopias Re-examined*

Modern Democracies

1776

"We hold these truths to be self-evident, that all men are created equal, that they are endowed by their Creator with certain unalienable Rights, that among these are Life, Liberty and the pursuit of Happiness. –That to secure these rights, Governments are instituted among Men, deriving their just powers from the consent of the governed." from The Declaration of Independence

In 1776, thirteen American colonies declared their independence from Britain and fought the War of Independence. Influenced by developments in Europe, especially France and England, the American founding fathers wrote a constitution for a democracy based on equal rights for all. Admittedly, "all" was initially something of an overstatement since it excluded women, non-landowners, Native Americans and Blacks. It took the Civil War in 1865 before slaves were even theoretically granted the rights of citizenship, and the constitutional amendment giving women the right to vote was not passed until 1920, less than a hundred years ago.

Yet it was government subject to a constitution. It provided for majority rule that also protected the rights of minorities insofar as they do not infringe on the common good. The state was separate from religion, and the purpose of the law was not to promote any single religious belief but religious tolerance. It remains a high standard.

Holocene Epoch   13,000 ya – present

Modern Life Emerges: *In the Last 600 Years*

more about...
the enlightenment

By the 18th century, democracy was an idea whose time had come. With Protestantism, Western Christianity had been fractured and Europe was torn by religious wars in which the king often arbitrarily determined the legality of religious practice. Philosophers of the Enlightenment began to think about a more rational society that could protect the rights of the individual from these arbitrary shifts of government. They came up with modern democracy.

• Paradoxically, the framers of the Constitution in the United States drew much of their inspiration from England, the country from which they had just declared independence. They believed they were fighting for freedoms that had been guaranteed in 1215 by the Magna Carta which placed England's sovereigns under the rule of law. In 1689, England itself had set up a constitutional government, transferring power from the monarchy to an elected Parliament. Today, a copy of the Magna Carta stands alongside the U.S. Constitution in Washington, D.C.

• In France, Enlightenment writers like Diderot, Rousseau, and Voltaire championed human rights and individual freedom. In 1789, the French revolution overthrew the absolute monarchy and reduced the control of the Roman Church. France experimented with a republic, an empire, and monarchy for 75 years but this first revolution permanently established the democratic principle and separated church from state.

Chapter 20: Industrial and Political Revolutions - *Utopias Re-examined*

The Victorian Age

1837

During the reign of Britain's Queen Victoria from 1837 to 1901, industrialization transformed landscapes everywhere. Rivers were re-routed, canals and bridges built, railways laid, massive tunnels dug for drains, subways and water delivery. The first oil wells and pipelines were put into operation, and cities across entire continents connected to electricity. Global trade burgeoned. Radio, telephone, and telegraph were revolutionizing communications, and Henry Ford's Model T car was about to liberate private transportation.

By the end of the Victorian age, European powers dominated the world. The United States of America had expanded from the Atlantic to the Pacific Coast, but it was the huge and commanding British and European empires on which the sun never set that looked unassailable. The British Empire alone contained 400 million people, one of five people in the world.

Few guessed that in less than half a century it would be in tatters.

Holocene Epoch   13,000 ya – present

Modern Life Emerges: *In the Last 600 Years*

more about…
the sweet and bitter fruits of industrialization

The 1800's were prosperous, and men of power and often good will thought the improvements they could bring to society were unlimited.

- The owners of many factories believed they had a duty of care toward their workers and provided them with housing, education, a church, cooperative stores and banks. As industrialization spread around the world, global trade made people rich and growing middle classes had fresh water, better food, hygiene, clothes, transportation, and opportunity. The lives of many were vastly better.

But industrialization also brought misery and injustice.

- Machinery pushed people off the farms into squalid cities to find work. Children were forced into hard labor, often into mine tunnels and chimneys where adults couldn't go. Global trade included slaves, the poor were often without hope, debtors were condemned to the workhouse, the sick had little access to medical care, and workers had few legal protections.

Responses in the face of moral outrage and social unrest were varied.

- In Europe, two systems emerged, each arguing it could serve the interests of society best. The archetypical struggle between capitalism and socialism that reached its climax in the 20th century was first given shape by Karl Marx, a German philosopher in England, who published *Das Kapital* in 1883.

- In America, a Civil War in which 600,000 people died – 1/6th of the population - resulted in the emancipation of the slaves, and a strong federal government. Workers organized into unions, and Congress passed laws protecting children and constraining unbridled capitalism.

Chapter 20: Industrial and Political Revolutions - *Utopias Re-examined*

Chapter 21

The Twentieth Century

1900-2000

Modern Life Emerges: *In the Last 600 Years*

Our Most Recent Legacies

Wars fought in the 20th century brought about the worst devastation produced by fighting in human memory.

But it was also a century in which human rights made giant steps. Women's right to vote, and the civil rights of minorities gave to millions of disenfranchised men and women the equal status inherent in a truly democratic government.

Although it was the century which invented the atomic bomb used as a weapon of war, it also introduced transforming technologies that changed people's lives for the better. Electricity, cars and air travel, the internet and space exploration all began during the 1900's

World Wars

1914 - 1945

For centuries, European nations had fought numerous wars over territory, trade, and religion. So when Britain, France and Russia became embroiled in another war in 1914 against Germany and Austria, it did not seem very different. But it was. New weapons supplied by industrialization brought horrors to this war more terrible than the world had ever seen. In the four horrifying years of World War I, trench warfare and tanks, aircraft and submarines, machine guns and mustard gas killed 10,000 a day, including 8 million troops and 6.5 million civilians. After the War, a world flu pandemic killed possibly 100 million more people.

World War II, from 1939 to 1945, grew out of issues unresolved by World War I. For six interminable years, an average of 210,000 people died each week in an orgy of brutality, torture, and starvation. Two-thirds of the people killed were non-combatants. Fourteen million died in German concentration camps and gas chambers, half solely because they were Jews. When it ended, more than 66 million people were dead. The world was left with the ashes of the atom bomb dropped on Japan by America, the awful potential of nuclear warfare, and a new world order.

Holocene Epoch   13,000 ya – present

Modern Life Emerges: *In the Last 600 Years*

more about…
20th Century Wars

In 1945, the world stood appalled before the almost unimaginable destruction and inhumanity wrought by the world wars. People determined it should never happen again.

• America's **Marshall Plan** helped rebuild Europe and Japan and helped establish flourishing economies and democracies there.
• The **United Nations** was set up to resolve international conflicts.
• European nations and Great Britain surrendered their colonial empires. Almost **40 new countries became independent** in less than thirty years.
• France and Germany joined in an Economic Union in the hope that economic interdependence would prevent another European war. Today the **European Union** consists of 27 nations stretching from Britain and Ireland to newly freed countries in eastern Europe.
• European domination was replaced by two superpowers wielding nuclear arms. America and Russia stared across the chasm of what was called a **Cold War** for the next 45 years. Sheer fear of the consequences of open hostilities was given the sanitized name of "mutual deterrence." It avoided full scale war for the rest of the century.

But these efforts did not eliminate war. Between 1945 and 1990, war has killed over 80 million people, more than died in both World Wars. Armed conflicts have broken out in China, Vietnam, Cambodia, Laos, Thailand, Korea, Indonesia, in much of Africa, Russia, Cuba, Argentina, the Falklands, El Salvador, Nicaragua and Guatemala, between India and Pakistan, between Greece, Cyprus and Turkey, and in many countries in Eastern Europe including most recently Bosnia and Yugoslavia. In Kuwait, Iran, Iraq, Lebanon, Israel, Palestine, and Syria, the Middle East has emerged as a major area of continued conflict.

In a remarkable process of statesmanship and reconciliation, South Africa moved from apartheid to democracy. But in much of the world, we have not yet learned how to solve our differences through negotiation and reconciliation.

Chapter 21: Twentieth Century - *Our Most Recent Legacies*

Human Rights

20th Century

If the 20th century saw perhaps the greatest and most horrifying abuses of basic human rights ever perpetrated by *Homo sapiens*, it also saw forceful action to protect those rights.

● The century opened with a strong movement demanding women's suffrage. Throughout the century, states around the world have given women the right to stand for election and to vote.

● The League of Nations was set up in 1919 following World War I. Its goal was to prevent another Great War by settling disputes through negotiation instead of through war. It was replaced by the United Nations after World War II, and in 1948 adopted the Universal Declaration of Human Rights. All human beings are born free with equal dignity and rights, which cannot be diminished by race, color, sex, religious or ethnic identity, social origin, birth, or nationality. The declaration includes the right to education, to work, and have a family. These principles underlie much of international law today.

● National liberation movements drove out many European colonial powers. Mahatma Gandhi led the movement of non-violence which freed India from British rule, and Nelsen Mandela headed the fight against apartheid in Africa.

● In the United States, the Civil Rights Movement between 1955 and 1968 led by Martin Luther King had a major influence in securing Black minorities rights which had theoretically been in place since Abraham Lincoln abolished slavery a century earlier.

● After the genocides of the wars in the Balkans and Rwanda, the United Nations set up the War Crimes Tribunal which continues today to indict and try those accused of crimes against humanity.

Holocene Epoch   13,000 ya – present

Modern Life Emerges: *In the Last 600 Years*

more about...
how the 20th century changed our lives

New technologies and inventions begun with the scientific revolution continued to accelerate during the 20th century. These inventions have not just been weapons of war but conveniences and communications systems that have transformed our lives. Many inventions have spread throughout the developed world so comprehensively that it is difficult to imagine how recent they are. Entire cities were connected to the electric grid, electric light replaced gas in our homes and electric appliances have taken over thousands of manual chores.

In Travel, the 20th century saw the first
- manned airplane, tractor, helicopter, Model T Ford, jeep
- tank, rocket, jet engine, turboprop engine, hovercraft, black box - flight recorder
- gas station, traffic signal, car radio, parking meter, and windshield wipers
- space travel, satellites, moon landing

In Communications, the 20th century saw the first
- radio, talking and Technicolor movies. photocopier, ballpoint pen
- zoom lens, stereo recording, tape recorder, television
- computer, mobile phones, the internet

In Medicine, 20th century developments include
- insulin, Valium, Prozac, Viagra, penicillin, oral contraceptives
- the iron lung, kidney dialysis machine, the internal pacemaker, artificial heart, contact lenses and silicone breast implants
- the elimination of small pox and vaccines for polio, measles, whooping cough, and HIV

In Business, notable innovations begun in the 20th century include
- bar codes and bar-code scanners, the ATM cash machine, credit cards and microchips
- McDonalds

Holocene Epoch   13,000 ya – present

Chapter 21: Twentieth Century - *Our Most Recent Legacies*

What Next?

The Third Millennium

With the collapse of the Soviet Union in 1991, a single superpower remained and many thought American democracy and capitalism would soon be embraced everywhere. Instead, the new millennium opened with "9/11," an assault striking at the heart of New York City, America's financial center, a searing demonstration that American values and her hegemony is actively opposed by many who are committed to different beliefs.

Countries like China, India, Russia, and Brazil are becoming powerful economic forces whose size is likely to surpass America's within decades. The world's use of fossil fuels is ballooning. Together we face new challenges of resolving opposing values and competing needs.

We don't know what will happen next, although we do know some of the questions. We know that part of what happens will depend on us. A history of arrogance, greed, stupidity, and cruelty might seem reason for despair.

But we also have a history of heroism, inventiveness, determination, and love that gives us reason for hope. The human species has a reservoir of good will, technological innovation, a capacity to work together, and a sense of justice that might yet save us.

Challenges and Possibilities

What will become of our amazing, probably unique, planet, its diversity, its mysteries, and its stunning beauty? What will become of us?

● What of our carelessness of our planet's resources? Could we destroy ourselves through environmental destruction and the species diversity on which our survival depends?

There is already evidence that we are responsible for a mass extinction of entire eco-systems that may rival the mass extinction that wiped out the mighty dinosaur which had survived 185 million years. We *Homo sapiens* have been here less than a quarter of a million years. Our survival is not guaranteed.

● Will we have enough food and water for the world's population? More than a billion people today do not have access to clean drinking water.

It is unlikely that we could grow enough food without the help of bees and other pollinating insects. Bee colonies have been collapsing, killing billions of bees worldwide. 90% of the wild bees in America have already disappeared for reasons that are not yet fully understood.

Can we solve this and problems like it?

● Humans seem willing to kill each other on a terrifying scale, dying both for trivialities and profound principle. Although we have waged war for as long as we can tell, our warfare has become increasingly lethal. We now destroy tens of millions of people in a single dispute, and with the proliferation of nuclear energy, we have the capacity to destroy the entire human race.

Will we find alternatives to killing people who worship a different God, who have values with which we disagree, who possess power or material riches we want for ourselves, or who may merely be desperate for enough food and water to sustain them?

There is reason both for hope and despair. Peoples have achieved justice and freedom without war. In our own times great men and women have fought without guns what often must have looked like overwhelming military power in India, in South Africa, in the United States, in Asia. But we are also a trigger-happy species often glorifying wars that are not just and are often self-destructive and fool-hearty.

The Third Millennium: What Next?

249

• What about energy? Will we be able to find safer, cleaner sources of energy that will not destroy the planet we call home?

Will we figure out how to provide these fundamental essentials to everyone?

• AIDS, SARS, and the Bird Flu have alerted us in recent years to the potential of pandemics to which our limited genetic diversity leaves us particularly vulnerable. Global travel spreads disease much faster now than at the time of the Black Death.

Will a pandemic happen again? Will science be able to limit its effects?

• Despite a run of luck lasting some fifteen thousand years, scientists assure us that our planet will experience what they ominously call "Catastrophic Global Geophysical Events" or, in their more playful moments, "gee-gees."

In recent years, we know that floods, earthquakes, tsunamis have killed hundreds of thousands of people in a single catastrophic blow.

Meteor and asteroid hits, solar rays and radiation from outer space, supervolcanoes, giant waves, and long ice ages are all inevitable.

Challenges and Possibilities

● In about four billion years, our sun will run out of energy. Earth, our solar system, and some day our galaxy will burn out. Ultimately, even the entire Universe will be unrecognizable. It may collapse again into the singularity out of which it originally exploded. Or it may continue to expand into infinity.

Are there other universes that could eventually take the place of the one we live in? Many scientists think there are. Scientists do know that energy is indestructible. In other words, energy will not cease to exist even if our entire universe no longer exists as a recognizable entity. Some theologians believe this energy is the undying, never-ending manifestation of what some might call "God."

Is there a supernatural universe like the one Plato suggested? Are there supernatural universes like heaven and hell that many Christians and Muslims believe in?

The whole story has not yet been told. We may live in hope.

We live in mystery.

The Third Millennium: What Next?

What Else to Read

The following list is a small selection of books written by scholars for readers who aren't necessarily scientists but want to know more.

Broadly speaking

A Short History of Nearly Everything, Bill Bryson (New York: Broadway Books, 2003). *With lucid humor, this book covers just about everything its title says it does. A wonderful read.*

Maps of Time: An Introduction to Big History, David Christian (Berkeley and Los Angeles: University of California Press, 2004). *A demanding but readable book by a leading academic, telling the story of time from the Big Bang to the twentieth century.*

About The Universe

Big Bang: The Most Important Scientific Discovery of All Time and Why You Need to Know About It, Simon Singh (New York: Fourth Estate, 2005) *If anyone can make the Big Bang intelligible to the layman, it is Singh. The book is an enlightening history of cosmology and the foibles of cosmologists.*

A Brief History of Time: The Updated and Expanded Tenth Anniversary Edition, Stephen Hawking (New York: Bantam Books, 1998). *This is a mind-stretcher but it's a classic and even if you understand only half of it, the excursion is simply incredible.*

The Elegant Universe, Brian Greene (New York: Norton, 1999). *A challenging attempt applying String Theory to resolve the contradictions and questions inherent in current models of the universe.*

Earth Before Humans

The Ancestor's Tale: A Pilgrimage to the Dawn of Life, Richard Dawkins (Houghton Mifflin, 2004). *A Chaucerian story of evolution, beginning with man and reversing to the microbe. Readable and relevant, written by an eminent evolutionist.*

Life: A Natural History of the First Four Billion Years of Life on Earth, Richard Fortey, (New York: Vintage, 1999). *A clear and accessible explanation of the key events of life on earth, from the first microbes to Cro-Magnon people.*

Ice Ages: Solving the Mystery, John Imbrie, Katherine Palmer Imbrie (Cambridge, Mass: Harvard University Press, 1986). *An explanation of why ice ages happen and Earth will continue to have them.*

The Earth: An Intimate History, Richard Fortey (Harper Collins, 2004). *The author is himself a geologist and his book turns the exploration of plate tectonics into a page-turner.*

Lives of the Planets: A Natural History of the Solar System, Richard Corfield, (Basic Books, 2007). A description of our solar system and what makes each of its planets both fascinating and unique.

After Humans Evolved

Ideas: A History from Fire to Freud: Peter Watson (London: Weidenfeld & Nicolson, 2005) *This is a huge book, over 800 pages long with pages almost half a square foot. But it is a fascinating read, and rewards the effort it asks. The Author's Note also includes a superb reading list of what Watson considers "indispensable."*

A Concise History of the World, J M. Roberts (New York: Oxford University Press, 1995). *A comprehensive overview of world history from the evolution of hominids to end of the twentieth century.*

The Neanderthal's Necklace: In Search of the First Thinkers, Juan Luis Arsuage (Chichester: Wiley, 2003). *A palaeo-anthropologist's work examining fossils left by humans in present-day Spain before Homo sapiens arrived. He asks why H. sapiens survived and H. neanderthalis did not.*

A Brief History of the Human Race, *Michael Cook (New York: Norton, 2003). An easy-to-read series of essays about human history on each of the world's continents.*

Out of Eden: The Peopling of the World, Stephen Oppenheimer (London: Constable & Robinson, 2003). *The evidence supporting the theory that all non-Africans in the world today are descendants of a pioneering group that left Africa 80,000 years ago.*

The Dawn of Human Culture, Richard G. Klein with Blake Edgar (New York: Wiley, 2002). *Klein begins with the bi-peds, and examines emerging evidence of human consciousness in Africa until 50,000 years ago when he believes a genetic change resulted in the final step in the evolution of modern thought.*

Since the Last Ice Age

After the Ice: A Global Human History 20,000-5000 BC, *Steven Mithen (London: Weidenfeld & Nicolson, 2003). An examination of cultures at the peak of the last ice age 20,000 years ago and the unique and dramatic changes which took place over the next 15,000 years with global warming. It covers changes in Asia, Europe, the Americas, Greater Australia, and Africa.*

Guns, Germs, and Steel: The Fates of Human Societies, Jared Diamond (New York: Norton, 1999). *A gripping analysis of human societies, examining why some communities gain ascendancy over others. It destroys comfortable theories that conquering cultures are more intelligent or morally superior.*

Cities, John Reader (William Heinemann, 2004). *The astonishing, ingenious and sometimes deplorable story of cities, from the first in Mesopotamia to our own metropolis creations.*

From Eve to Dawn, Marilyn French (McArthur, 2003). *A three-volume work examining the role of women in cultures around the world from prehistory to now. It's a sobering historical analysis, written by a refreshingly optimistic and extraordinarily well-informed, author.*

Civilization One: The World is Not as You Thought It Was, Christopher Knight, Alan Butler (Watkins/Duncan Baird Publishers, 2004). *A provocative analysis of stone monuments built 4-6,000 years ago. The builders had sophisticated astrological skills, and their numbers are still evident in our own weights and measurements.*

1491: New Revelations of the Americas Before Columbus, C (Vintage Books, 2006) *This is a book about some of the more recent and quite astonishing discoveries about life and civilizations in the Americas before the arrival of Europeans in 1492.*

Recent History of Politics, Religion, and Science

Science: A History 1543-2001, John Gribbin (London: Penguin Books, 2002). *A detailed chronicle of major scientific achievements from the Renaissance to today's exploration of space.*

The Reformation: A History, Diarmaid MacCulloch (New York: Viking Penguin, 2004). *A definitive study of the tortured, complex, and evolving relationships and struggles between religion, politics, and science from 1490 to the time of modern democracies.*

The Modern Mind: An Intellectual History of the Twentieth Century, Peter Watson (New York: Harper Collins, 2000). *This is another 800+ page book chronicling the immense power of ideas that shaped our modern world. Reading every word requires committed interest but it is equally valuable as a reference book.*

On the Internet

The internet is increasingly one of the best places for non-academics to stay in contact with the newest research, discoveries, and thinking about our Universe and life within it.

www.Nasa.gov
The home page can access sites on the solar system, the universe, and earth.

www.nasa.gov/astrophysics
A gateway to the big questions - and some of the answers - scientists are trying to answer today.

www.NewScientist.com
A popular and readable magazine also available on line featuring some of the arresting findings by science in every field.

www.ucmp.berkeley.edu/exhibits/index.php
University sites on understanding science, the history of life on earth, geological times, evolution, and fossils.

www.AllOfTimeOnLine.com
www.TheBigBangToNow.wordpress.com
Companion website and blog to this book.

News sites and science blogs are written in accessible language and are among popular sites that have the latest research and reports from scientific conferences around the world.

www.BBC.co.uk/science

www.Telegraph.co.uk/science

www.NYTimes.com/science

www.HuffingtonPost.com/news/science

www.economist.com/science-technology

Appendices

Although this book is written mostly in informal English, a few Greek translations, geological and archaeological time scales, and modern names for ancient places on the next pages may help you navigate around if you wish to read more in other sources.

Translations from the Greek

Translations of Greek terms commonly used to name many geological and archaeological times can them easier to understand.

GREEK TERM..........ENGLISH TRANSLATION

     paleo......................oldest

     meso......................middle

     neo..........................new

     protero....................first

     phanero...................visible

     -cene.......................recent

     -zoic........................life

     -lithic.......................stone

Geological Time Scale

Time on Earth is divided by geologists into 4 great Eons. The last two eons are divided into Eras, which are divided into Periods. The two most recent periods are further divided into Epochs.

Geological times are based on findings of rock and early fossils by geologists and palaeontologists and are the best estimates we have of what happened when. Most estimates are approximate and change as new evidence is found and techniques for analyzing it improve. Many times, we discover that organisms appeared much earlier than we first thought.

It may also be helpful to know there is some disagreement among scholars about these timescales. Dates for different time frames sometimes vary among experts by millions of years, names for different times are not nearly as neat as one might expect and there is disagreement over whether various times constitute eons, eras, periods or epochs. American and European names do not always match, and textbooks use whatever system the author likes best. The Cenozoic Era has a Quaternary (fourth) period. There used to be first, second and third periods, too, but they were replaced by the Paleogene and Neogene periods. Some scientists still refer to the Tertiary and Quaternary periods.

For the non-professional, it isn't important to know this. What matters is that there is broad agreement on the order in which developments occur.

There is, as with almost everything we know, room for disagreement.

Geological Ages

HADEAN EON: AGE OF METEORITES 4.55 BYA
Hell fire and volcanoes; the moon; earliest rocks, water and atmosphere form

ARCHEAN EON: *EARLIEST ANCIENT TIME* 4.00 BYA
Earliest microscopic life; major formation of Earth's crust

PROTEROZOIC EON: OLDEST LIFE IS ESTABLISHED 2.50 BYA
Oxygen increases; life becomes more complex

    PALEOPROTEROZOIC ERA: OLDEST LIFE 2.5 BYA
    Ends with oxygen poisoning extinctions

    MESOPROTEROZOIC ERA: MIDDLE OLDEST LIFE 1.7 BYA
    Eukaryotes, single-cell plants and proto- animals develop; atmosphere begins to become oxygenated

    NEOPROTEROZOIC ERA: NEW OLDEST LIFE 1.0 BYA
    Earliest multi-celled organisms

        TONIAN PERIOD 1 BYA
        first multi-celled plants and animals

        CRYOGENIAN PERIOD *COLD TIME* 850 MYA
        intense glaciations; possibly Earth's coldest period

        EDIACARAN PERIOD *900* MYA
        multitudes of multi-celled organisms some with shells

PHANEROZOIC EON: VISIBLE LIFE 542 MYA
Evolution of animals; life moves onto land; evolution of humans

PALEOZOIC ERA: OLD (VISIBLE) LIFE 542 MYA
Time of marine animals

CAMBRIAN PERIOD 543 MYA
evolutionary explosion; all familiar animal groups develop

ORDOVICIAN PERIOD 488 MYA
plants spread to land; earliest vertebrate fish; mass extinctions

SILURIAN PERIOD 444 MYA
first land-dwelling anthropoids

DEVONIAN PERIOD 416 MYA
mountain ranges develop; amphibians evolve; mass extinctions

CARBONIFEROUS PERIOD 359 MYA
coal deposits; winged insects; amniotic egg

PERMIAN PERIOD 299 MYA
the great dying extinction

MESOZOIC ERA: MIDDLE LIFE 251 MYA
Age of reptiles

TRIASSIC PERIOD 250 MYA
reptiles diversify; dinosaurs and earliest mammals evolve

JURASSIC PERIOD 200 MYA
dinosaurs dominate earth; birds evolve

CRETACEOUS PERIOD 145 MYA
flowering plants evolve; chalk deposits; placental mammals

CENOZOIC ERA: RECENT LIFE 65.5 MYA
Age of mammals

PALEOGENE PERIOD 65.5 MYA
the first primates and prairie grasses

PALEOCENE EPOCH 65.5 MYA
new mammals including first primates

EOCENE EPOCH 55.8 MYA
first apes

OLIGOCENE EPOCH 33.9 MYA
long-term cooling; new grasses

NEOGENE PERIOD 23 MYA
description

MIOCENE EPOCH 23 MYA
hominoids evolve

PLIOCENE EPOCH 5.3 MYA
stone tools appear

QUARTERNARY PERIOD 2.59 MYA
magnetic reversal and climate changes

PLEISTOCENE EPOCH 2.59 MYA
periods of glaciation: "the ice age"; Homo genus evolves

LOWER PLEISTOCENE 1.8 MYA
homo ergaster and homo erectus evolve

MIDDLE PLIESTOCENE 780,000 YA
homo sapiens evolves in Africa

UPPER PLEISTOCENE 127,000 YA
homo neanderthal evolves in europe

HOLOCENE EPOCH 13,000 YA
current age with warm interglacial; first civilizations

Archaeological Ages

Although the geological time scale is used for the earliest period of Earth history, scientists from different fields name periods of time differently. Archaeologists and palaeontologists studying the evolution of cultures have stages based on the kind of tools that were used. Different tools occur at different times in different geographical regions and not all cultures developed them all. When ages aren't fixed for everyone everywhere at the same time, archaeological ages are used instead of fixed geological times.

- **Palaeolithic Stone Age:** The Old Stone Age is the earliest known stone age and first occurred about 2.5 million years ago in Africa.

- **Mesolithic Stone Age:** The Middle Stone Age is between the Old and New Stone Ages. It isn't very well defined and is used when a culture doesn't seem to fit clearly into either earlier or later stone ages.

- **Neolithic Stone Age:** The earliest Neolithic or New Stone Age began about 10,000 years ago when farming started. There are still societies whose tools are made of stone rather than of metal who, by this definition, are living in a Neolithic stone age today.

- **Copper Age:** The first known society to move into the Copper Age was in the Middle East about 7,000 years ago when metal tools began to be used.

- **Bronze Age:** Bronze is made by smelting copper and tin together. It often, but not always, follows a copper age. Bronze tools first appeared in the Middle East about 5,600 years ago.

- **Iron Age:** The first known Iron Age followed the Bronze Age about 2,500 years ago.

Modern Names of Ancient Places

Between 9,000 and 3,000 years ago, many names were given to places in the area called the Cradle of Civilization, where the earliest known civilizations began in the Middle East. These lands, embraced by the Black and Caspian Seas, the Persian Gulf and the Arabian and Mediterranean Seas, today include Iraq, Iran, Turkey, Lebanon, Syria, Jordan, Israel and Palestine.

IN MESOPOTAMIA, TODAY'S IRAQ:

- Mesopotamia, the Fertile Crescent, the area in Iraq between the Euphrates and Tigris Rivers
- Jarmo, one of the world's earliest cities, located in northern Iraq
- Sumer, (given the Biblical name "Shinar") in the Tigris-Euphrates valley; Kish, Lagash, Eridu, Ur and Uruk were also prominent Sumerian cities
- Akkadia, with its capital Agade eventually included Sumer and much of what today we call the Middle East
- Babylonia, the capital of a dynasty founded by the Amorites. At one time it controlled most of Mesopotamia

IN PERSIA, TODAY'S IRAN:

- Elam Kingdom, Median Empire, the several Persian Empires and Seleucid Empire

ANATOLIA, TODAY'S TURKEY:

- The cities of Catalhuyuk, and later Troy were built here
- Hittite Empire began here, spreading through most of Asia Minor

IN MODERN LEBANON:

- Canaan, later called Phoenicia, the seat of far-flung sea-trade
- Jericho, located on the Jordan River, one of the earliest cities

263

What Scholars Study

Scientists and scholars in many fields contribute to the study of our past. Most specialize within their field as well as cooperate in joint studies with other scientists in other fields.

ANTHROPOLOGISTS study contemporary cultures, often by participant observation of a people's social institutions, myths, religion, and customs.

ARCHAEOLOGISTS study ancient peoples and cultures by examining and often excavating ancient artifacts, structures, and other remains.

ASTRONOMERS study bodies like planets, stars, meteors, galaxies, and black holes in outer space.

CLIMATOLOGISTS are meteorologists who specialize in studying climate change and its effects on global systems and living organisms.

COSMOLOGISTS are usually physicists and astronomers working to understand the principles of the universe, how it originated and continues to operate.

FORENSIC SCIENTISTS analyze biological, chemical, and physical samples to provide evidence for what has taken place at a sight. Many forensic scientists work with the police to investigate crime scenes, but their skills are also invaluable in examining sights which may be thousands or even millions of years old.

GENETICISTS examine generational and species DNA; they trace changes in DNA to understand our past and the evolution of life.

GEOLOGISTS study the formation of Earth's crust and the movement of tectonic plates, especially through the examination of rocks.

HISTORIANS are usually scholars who study recorded or written events; time before writing is called pre-history, which is more often studied by palaeontologists and archaeologists.

LINGUISTS study the structure of language; historical, comparative, and geographical linguistics is often helpful in understanding relationships and movement of peoples.

MATHEMATICIANS operate in almost every area of science. They examine data to calculate the probability that some events such as climate change or asteroid hits will happen. They estimate the chances that some recent or long-ago event or result happened by chance. Like Newton or Einstein, they develop theories to bring together what may otherwise appear to be a confusing array of unconnected or contradictory facts,

METEOROLOGISTS work to understand and forecast weather and climate change.

PALAEONTOLOGISTS study pre-historic forms of life preserved in fossils, the remains of ancient plants and animals.

PHYSICISTS study the basic properties of matter and energy like nuclear or thermodynamic energy.

VOLCANOLOGISTS study volcanoes, frequently by visiting active sites, which makes it one of the more dangerous scientific professions.

Index

dinosaurs, 16, 17, 63, 80,
81, 83, 84, 85, 86, 87,
88, 89, 90, 91, 92, 97,
100, 102, 103, 104, 105,
108, 110, 111, 260

E

egg, 58, 67, 74, 75, 95,
101, 260
Egypt, 175, 177
embryo, 67, 75
empire, 183, 187, 190, 191,
196, 198, 201, 235, 237
energy, 22, 23, 26, 41, 44,
45, 48, 49, 53, 94, 107,
111, 130, 131, 197, 231,
234, 249, 250, 251, 265
Enlightenment, 237
eukaryote, 53
evolution, 15, 41, 47, 54,
55, 63, 77, 80, 118, 124,
125, 139, 148, 227, 253,
254, 256, 260, 262, 264
extinction, 18, 41, 47, 48,
49, 54, 59, 62, 63, 70,
71, 78, 81, 86, 87, 89,
102, 104, 105, 143, 150,
158, 248, 260

F

farming, 17, 114, 160, 161,
163, 166, 167, 168, 170,
181, 182, 231, 235, 262
Fertile Crescent, 172, 263
fire, 34, 69, 104, 126, 130,
131, 171, 207, 259
flatworm, 57, 69
flowers, 201

fossils, 42, 54, 55, 57, 69,
83, 85, 90, 100, 121,
132, 138, 141, 142, 148,
149, 153, 254, 256, 258,
265
fruiting plants, 107
fungus, 60

G

galaxies, 23, 26, 28, 29, 33,
264
Gauls, 177, 197
glacial, 64, 77, 115, 125,
130, 132, 152, 160, 163,
164
global warming, 58, 63, 79,
115, 226, 231, 254
Gondwana, 62, 113
Greeks, 93, 173, 185, 186,
217
gunpowder, 205

H

Hebrew, 178, 179
Hinduism, 189
Hippocrates, 187
hominid, 119
hominin, 128, 158
hominoids, 117, 118, 119,
261
homo, 119, 261
human, 21, 35, 63, 102,
105, 117, 118, 121, 123,
124, 125, 126, 127, 128,
129, 130, 131, 133, 136,
138, 139, 141, 142, 144,
145, 147, 148, 149, 150,
152, 154, 155, 157, 159,
160, 162, 164, 165, 174,

181, 182, 185, 195, 200,
211, 213, 217, 229, 231,
233, 237, 241, 244, 247,
249, 254

I

ice ages, 58, 59, 71, 77,
  156, 226, 231, 250, 253
immune system, 65, 69
Incas, 183
India, 62, 112, 113, 141,
  156, 175, 176, 177, 188,
  189, 201, 207, 214, 215,
  224, 243, 244, 247, 249
insects, 57, 64, 80, 88, 89,
  91, 92, 93, 95, 249, 260
interglacial, 77, 115, 125,
  161, 163, 164, 165, 231,
  261
Islam, 179, 185, 199, 200,
  201

J

jellyfish, 55

L

Laurasia, 113
light year, 29, 35
Local Bubble, 124
Lucy, 121

M

Magna Carta, 237
magnetic, 37, 124, 205,
  214, 261
mammals, 57, 63, 80, 81,
  83, 84, 85, 86, 87, 89,
  91, 94, 97, 100, 101,

102, 104, 105, 106, 107,
  108, 109, 111, 114, 119,
  159, 260, 261
matter, 22, 24, 25, 35, 43,
  48, 187, 265
Mayan, 183
measles, 65, 245
Mesopotamia, 172, 173,
  254, 263
metallurgy, 170, 171
meteor, 79, 96, 102, 103
methane, 33, 34, 102
migration, 64, 145
Milky Way, 19, 28, 29, 32,
  33, 35
Minoan, 177
monasteries, 196, 213, 221
moon, 31, 33, 36, 37, 60,
  61, 140, 173, 245, 259
mountains, 39, 47, 66, 71,
  112, 168
mouse, 100, 101, 111
multicellular, 50, 55, 56, 63
mumps, 65, 225
mystery, 177, 227, 251

N

Neanderthals, 142, 143,
  147, 148, 149
nectar, 95
Norsemen, 197, 199

O

ocean, 36, 39, 42, 47, 71,
  103, 159, 231
Oceania, 162, 180
oil, 98, 99, 238
organism, 43, 55, 111

supervolcano, 79, 145, 150, 151, 152
survival of the fittest, 87

T

technology, 69, 126, 131, 170, 211, 216, 256
tectonic, 39, 46, 47, 98, 151, 265
termites, 92
tetrapods, 69, 111
textiles, 168
tools, 122, 123, 128, 130, 131, 139, 141, 153, 156, 158, 170, 171, 172, 261, 262

U

universe, 15, 20, 29, 43, 189, 229, 251, 252, 256, 264

V

vascular tubes, 67
vertebrate, 57, 111, 260
Vikings, 197
volcano, 151

W

war, 71, 160, 167, 170, 175, 179, 185, 192, 205, 207, 209, 211, 222, 223, 233, 241, 242, 243, 244, 245, 249

Cut out this page and use it as a bookmark.

Where are we now?

 Five Big Start-Ups
> The Universe - 13.7 billion years ago
> Earth - 4 ½ billion years ago
> 1st life on Earth - 4 billion years ago
> True plants & animals - 1.7 billion years ago
> 1st land life - 470 million years ago

 The Dinosaurs
> 1st Reptiles, dinosaurs, & mammals - 250 million years ago
> Dinosaurs' extinction - 65 million years ago

 Age of Mammals
> 1st primates - 60 million years ago
> Primates walk on 2 feet - 7 million years ago
> 1st humans - 1.9 million years ago

 First Age of Homo Sapiens
> 1st Homo sapiens - 250,000 years ago
> Homo sapiens leaves Africa - at least 80,000 years ago

 Farming, Cities and Civilizations
> Beginning of settled farming - 11 thousand years ago
> Greek democracy; 1st great religions - 2.7 thousand years ago
> Christianity as a major political power - dark and middle ages

✏ **Modern Life Emerges - last 600 years**
> Europe's global explorations, printing press
> Protestant Reformation, religious wars
> Scientific Revolution
> Industrial Age, modern democracies
> World Wars, space age

Reminder

Just as a thousand years is a thousand times longer than one year...
- A million years is a thousand times longer than a thousand years.
- A billion years is a thousand times longer than a million years.

35016152R00152

Made in the USA
Middletown, DE
31 January 2019